The 100 Strangest Things in the World

STEVEN CURRY

The 100 Strangest Things in the World

Author: Steven Curry
Publisher: L'Oliveto Publisher

ISBN 9798345384923

© 2024 Steven Curry
© 2024 L'Oliveto Publisher
First print edition November 2024

L'Oliveto Publisher
Rome, Italy
www.oltreilcespuglio.it

This colophon certifies that this book was published by L'Oliveto Publisher and represents a quality production made with care and dedication. We are committed to promoting the spread of knowledge and culture through our publications.
For information on other titles published by L'Oliveto Publisher, visit our website or contact us directly at the e-mail address or phone number provided above.

Thank you for choosing L'Oliveto Publisher. Happy reading!
"The 100 Strangest Things in the World"

INDICE

5

Chapter 1: Strange Creatures

1. Axolotl

The axolotl, often called the Mexican walking fish, is a unique creature native to the lakes underlying Mexico City, especially Lake Xochimilco. Despite its fish-like appearance, the axolotl is not a fish but an amphibian, specifically a type of salamander. What makes the axolotl particularly strange and fascinating is its ability to regenerate entire limbs, spinal cord segments, and even parts of its heart and brain. This remarkable regenerative capability has made the axolotl a subject of intense scientific study, as understanding the mechanisms behind it could have profound implications for human medicine.

Axolotls exhibit neoteny, meaning they retain juvenile features throughout their adult lives. Unlike most amphibians that undergo metamorphosis, axolotls retain their larval features, such as gills, and live permanently in water. This retention of juvenile traits is one reason they look so peculiar, with their feathery external gills and wide, perpetual smiles.

Their diet in the wild consists of small prey like worms, insects, and small fish. In captivity, they are often fed a diet of bloodworms, brine shrimp, and commercial fish pellets. Axolotls are known for their docile nature and relatively simple care requirements, which makes them popular in the pet trade. However, their natural habitat is under threat due to urbanization, pollution, and the introduction of invasive species, making the axolotl critically endangered in the wild.

Despite their endangered status, axolotls are bred extensively in captivity for research and the pet trade. This breeding has led to a variety of color mutations, including leucistic (white with pink gills), albino, golden albino, and melanistic (all black) axolotls.

11

These color variations add to their appeal and curiosity among pet enthusiasts and researchers alike.

Axolotls have an interesting cultural significance in Mexico. The Aztecs revered them, naming them after the god Xolotl, who was associated with fire and lightning and was believed to guide the souls of the dead to the underworld. According to legend, Xolotl transformed into an axolotl to escape sacrifice, which ties into the creature's regenerative abilities and its somewhat otherworldly appearance.

In the realm of scientific research, axolotls are invaluable. Their ability to regenerate tissues without scarring has implications for regenerative medicine and understanding how to promote healing in humans. Researchers are studying the genetic and molecular basis of this regeneration, hoping to unlock secrets that could lead to breakthroughs in treating injuries and diseases.

The axolotl's capacity for regeneration, coupled with its neotenous nature and endangered status, makes it one of the strangest and most intriguing creatures on our planet. Its perpetual state of youth and remarkable healing abilities continue to captivate scientists and animal lovers worldwide, ensuring that the axolotl remains a subject of fascination and hope for future scientific advancements.

2. Blobfish

The blobfish, often referred to as the world's ugliest fish, inhabits the deep waters off the coasts of Australia and Tasmania. Its gelatinous appearance has made it an icon of the strange and unusual in the marine world. The blobfish's unique look is a result of its adaptation to the extreme pressure of the deep sea, where it resides at depths ranging from 600 to 1,200 meters.

Unlike most fish, the blobfish lacks a swim bladder, the gas-filled organ that helps fish maintain buoyancy. Instead, its body is composed of a gelatinous mass slightly less dense than water. This allows the blobfish to float above the sea floor without expending energy. This adaptation is essential for survival in its high-pressure, low-energy environment, where food is scarce and movement must be minimized to conserve energy.

The blobfish's diet consists mainly of small crustaceans and other invertebrates. It feeds by simply opening its mouth and letting its prey drift in, a passive feeding strategy that suits its energy-conserving lifestyle. Despite its unappealing appearance, the blobfish plays a crucial role in its ecosystem, contributing to the balance of marine life at extreme depths.

In its natural habitat, the blobfish looks quite different from the sad, droopy-faced creature often depicted in photographs taken at the surface. When removed from the high-pressure environment of the deep sea, its gelatinous body expands and deforms, resulting in the comically miserable expression that has earned it a spot in various "ugly animal" competitions. In its natural setting, however, the blobfish is well-adapted and functions perfectly within its ecological niche.

The blobfish's notoriety has sparked conversations about the conservation of deep-sea habitats. As fishing practices and pollution increasingly threaten these environments, the blobfish has become a symbol of the unique and often misunderstood creatures that reside in the deep ocean. Efforts to protect these habitats are crucial, not only for the survival of the blobfish but also for maintaining the biodiversity of our planet's oceans.

Despite its unfortunate appearance and the challenges, it faces in its natural habitat, the blobfish is a resilient species. It has adapted to one of the most extreme environments on Earth, demonstrating the incredible diversity of life forms and survival strategies in the animal kingdom. The blobfish's story is a reminder of the wonders of nature and the importance of preserving even the most unusual and unglamorous creatures.

As a subject of scientific curiosity and conservation efforts, the blobfish has gained a surprising amount of attention and even a kind of celebrity status. It has become a mascot for various conservation campaigns and educational programs, highlighting the importance of protecting marine life. The blobfish's unlikely charm lies in its ability to captivate human interest, drawing attention to the often-overlooked depths of the ocean and the remarkable creatures that inhabit it.

In summary, the blobfish, with its unique adaptations and unfortunate surface appearance, stands as a testament to the wonders of evolution and the diversity of life on Earth. Its story underscores the importance of understanding and protecting the fragile ecosystems of the deep sea, ensuring that even the strangest and least glamorous creatures have a place in our world.

3. Platypus

The platypus, native to eastern Australia and Tasmania, is one of the most peculiar creatures in the animal kingdom. This semi-aquatic mammal, part of the monotreme order, lays eggs instead of giving birth to live young, a trait that sets it apart from most other mammals. The platypus's unique blend of characteristics from different animal classes has fascinated scientists since its discovery. One of the most striking features of the platypus is its duck-like bill, which is not only used for feeding but also equipped with electroreceptors. These receptors allow the platypus to detect electrical signals produced by the muscles and nerves of its prey, making it an adept hunter even in murky waters where visibility is low. This adaptation is particularly useful for finding food such as insects, larvae, shellfish, and worms.

The body of the platypus is covered in dense, waterproof fur that provides insulation in cold water. Its webbed feet make it an excellent swimmer, while its streamlined body allows it to glide effortlessly through the water. On land, the webbing on its feet retracts, exposing claws that help the platypus to walk and dig burrows. This versatility enables the platypus to thrive in both aquatic and terrestrial environments.

Another intriguing aspect of the platypus is its method of reproduction. As one of the few egg-laying mammals, the female platypus lays one to three eggs and incubates them by curling around them. After about ten days, the eggs hatch, and the mother nurses the young with milk secreted from mammary glands, though she lacks nipples. Instead, the milk is absorbed through the skin of the young.

Male platypuses possess a venomous spur on their hind legs, which can deliver a painful sting to potential predators or rivals. This venom is strong enough to incapacitate small animals and cause severe pain in humans. The presence of venom in the platypus adds another layer of intrigue, as it is one of the few venomous mammals.

The discovery of the platypus baffled early European naturalists. When the first specimens were sent back to England in the late 18th century, many scientists believed the animal to be a hoax, a combination of parts from different animals sewn together. The seemingly improbable combination of traits—such as a beak, webbed feet, and the ability to lay eggs—challenged existing classifications and led to significant debates in the scientific community.

The platypus's role in Aboriginal Australian culture is significant, with various tribes having their own stories and legends about this unique creature. It is often seen as a symbol of the diverse and extraordinary nature of the Australian wildlife.

In recent years, the platypus has become a symbol of conservation efforts in Australia. Habitat destruction, pollution, and climate change pose significant threats to its survival. Conservation programs aim to protect the natural habitats of the platypus, ensuring the survival of this enigmatic species.

Research into the platypus's genome has revealed fascinating insights into its evolutionary history and biology. The combination of reptilian, avian, and mammalian genes reflects its ancient lineage and provides valuable information about the evolutionary processes that have shaped modern mammals.

In conclusion, the platypus stands as one of the strangest and most remarkable animals on Earth. Its unique blend of characteristics, from egg-laying to venomous spurs, continues to captivate scientists and the public alike. As a symbol of Australia's rich biodiversity, the platypus reminds us of the wonders of evolution and the importance of preserving our natural world.

4. Narwhal

The narwhal, often dubbed the "unicorn of the sea," is a remarkable Arctic marine mammal known for its long, spiral tusk. This tusk, which can grow up to ten feet in length, is actually an elongated upper left canine tooth. Male narwhals typically possess these impressive tusks, although a small percentage of females also grow them. The function of the tusk has been a subject of scientific curiosity for centuries.

Narwhals inhabit the icy waters of the Arctic Ocean, particularly around Greenland, Canada, and Russia. They are well-adapted to their cold environment, with a thick layer of blubber that provides insulation against the frigid temperatures. Narwhals are deep divers, capable of reaching depths of up to 1,500 meters (nearly 5,000 feet) in search of food. Their diet mainly consists of fish, squid, and shrimp, which they hunt using echolocation.

The narwhal's tusk has fascinated humans for centuries, leading to numerous myths and legends. During the Middle Ages, narwhal tusks were often sold as unicorn horns, believed to possess magical properties and used as a status symbol among European nobility. These tusks were sometimes crafted into elaborate drinking horns and other ornamental items. The true nature of the narwhal and its tusk remained a mystery until more recent scientific studies provided insights into its biology.

Recent research has suggested that the tusk might serve multiple purposes. One theory is that it acts as a sensory organ, helping narwhals detect changes in their environment, such as variations in water temperature, pressure, and salinity. The tusk contains millions of nerve endings that connect to the narwhal's brain, supporting the idea of it being a sophisticated sensory tool. Another theory posits that the tusk plays a role in mating rituals and social dominance among males, as they are often observed engaging in tusk jousting behaviors.

Narwhals are social animals, typically found in pods of up to 20 individuals, though larger groups can form during migrations or in areas with abundant food. They communicate using a range of vocalizations, including clicks, whistles, and knocks, which are essential for navigation, hunting, and social interactions. These vocalizations are an important aspect of their echolocation abilities, allowing them to navigate the dark, icy waters of their habitat.

The narwhal's unique appearance and behaviors have made it a subject of extensive scientific study and conservation efforts. The International Union for Conservation of Nature (IUCN) currently lists the narwhal as Near Threatened, with climate change and hunting posing significant threats to their population. As sea ice continues to melt due to global warming, the narwhal's habitat is rapidly changing, impacting their migratory patterns and access to prey.

Traditional hunting by indigenous communities in the Arctic, who rely on narwhals for sustenance and cultural practices, is conducted sustainably. However, increased commercial hunting and illegal poaching for tusks have raised concerns about the long-term survival of the species. Conservation organizations are working to address these threats by promoting sustainable practices and protecting critical habitats.

The narwhal's tusk, once shrouded in mystery and myth, is now understood as a remarkable example of nature's ingenuity. As research continues, we gain deeper insights into the life and ecology of these elusive creatures, highlighting the importance of preserving their natural environment. The narwhal remains a symbol of the Arctic's unique biodiversity and the ongoing efforts to protect it.

In conclusion, the narwhal's distinctive tusk and its adaptations to the harsh Arctic environment make it one of the strangest and most fascinating creatures in the world. Its story intertwines mythology, scientific discovery, and conservation, underscoring the need to protect the delicate ecosystems that support such extraordinary wildlife.

5. Yeti Crab

The yeti crab, formally known as Kiwa hirsuta, is a fascinating and mysterious creature discovered in 2005 near hydrothermal vents in the South Pacific Ocean. This deep-sea crustacean is named for its furry appearance, with long, blond setae (hair-like structures) covering its pincers and legs, reminiscent of the mythical Yeti.

The yeti crab inhabits one of the most extreme environments on Earth, living around hydrothermal vents at depths of over 2,200 meters (7,200 feet). These vents emit superheated, mineral-rich water that can reach temperatures up to 400 degrees Celsius (752 degrees Fahrenheit). Despite these harsh conditions, the yeti crab has adapted remarkably well, thriving in an environment where few other creatures can survive.

One of the most intriguing aspects of the yeti crab is its unique symbiotic relationship with chemosynthetic bacteria. The bacteria live on the crab's hairy pincers, feeding on the chemicals emitted by the hydrothermal vents. In return, the yeti crab "farms" the bacteria by waving its pincers in the nutrient-rich water, facilitating the growth of the bacterial colonies. The crab then harvests and consumes the bacteria, making it an integral part of its diet. This symbiotic relationship is a prime example of the complex and often surprising interactions that can occur in nature's most extreme environments.

The yeti crab's appearance and behavior have captivated scientists and the public alike. Its setae-covered pincers are not just for farming bacteria but also serve a sensory function, helping the crab navigate the dark, deep-sea environment. These hairy structures are covered in chemical receptors that allow the yeti crab to detect the presence of other animals, food sources, and potential mates.

The discovery of the yeti crab has shed light on the incredible biodiversity found around hydrothermal vents, which are among the least explored ecosystems on Earth. These vents host a wide variety of unique species that have evolved to thrive in extreme conditions, relying on chemosynthesis rather than photosynthesis as the primary source of energy. The yeti crab is one of many remarkable organisms that contribute to the rich tapestry of life in these deep-sea habitats.

Kiwa hirsuta is not the only species of yeti crab discovered. Since its initial discovery, scientists have identified additional species within the Kiwa genus, including Kiwa puravida, found off the coast of Costa Rica, and Kiwa tyleri, discovered near hydrothermal vents in the Antarctic. Each of these species exhibits its own unique adaptations to the specific conditions of its environment, further highlighting the diversity and adaptability of life in the deep sea.

19

The yeti crab's discovery has sparked interest in deep-sea exploration and research, highlighting the importance of these unique ecosystems. Hydrothermal vent communities are not only biologically diverse but also geologically and chemically distinct, offering insights into the origins of life on Earth and the potential for life on other planets. Studying these environments helps scientists understand how life can thrive in extreme conditions, which has implications for the search for extraterrestrial life.

Conservation of deep-sea habitats, including hydrothermal vent ecosystems, is becoming increasingly important as human activities such as deep-sea mining and climate change pose significant threats. Protecting these fragile environments ensures the survival of the yeti crab and other unique species that depend on these habitats for survival.

In conclusion, the yeti crab stands out as one of the strangest and most remarkable creatures in the deep sea. Its furry appearance, symbiotic relationship with bacteria, and ability to thrive in extreme environments make it a symbol of the incredible adaptability and diversity of life on our planet. The continued study and conservation of deep-sea ecosystems are crucial for understanding and preserving the rich biodiversity that exists in these hidden corners of our world.

6. Star-Nosed Mole

The star-nosed mole (Condylura cristata) is one of the most unusual and specialized mammals in the world. Native to the wetlands and moist, lowland areas of eastern North America, this small mole is instantly recognizable by the unique, star-shaped structure on its snout. This extraordinary feature, composed of 22 fleshy appendages called tentacles, gives the star-nosed mole its name and serves as a highly specialized sensory organ.

These tentacles are covered with over 25,000 tiny sensory receptors known as Eimer's organs. These receptors allow the star-nosed mole to detect minute vibrations and textures, making it one of the most sensitive touch organs found in the animal kingdom. The star-shaped nose is so efficient that it can identify and consume prey in as little as 120 milliseconds, making the star-nosed mole the fastest-eating mammal on the planet. This incredible speed is vital for survival in the mole's wetland habitat, where food competition is fierce.

The star-nosed mole is primarily carnivorous, feeding on small invertebrates, insects, worms, and aquatic organisms. It is an excellent swimmer and often forages for food underwater, using its star-shaped nose to explore and capture prey. The mole's unique hunting technique involves rapid tapping movements with its tentacles, allowing it to quickly scan and identify potential food items. This rapid tactile feedback system enables the mole to thrive in its often dark and murky environment.

In addition to its remarkable sensory abilities, the star-nosed mole has several other adaptations suited to its wetland habitat. It possesses large, paddle-like forelimbs with strong claws for digging through soil and creating extensive tunnel systems. These tunnels provide shelter, breeding grounds, and a means of locating food. The mole's fur is water-repellent, helping it stay dry and insulated while swimming or burrowing in damp conditions.

The star-nosed mole's eyes are small and poorly developed, as vision is not crucial in its subterranean and aquatic environments. Instead, the mole relies heavily on its sense of touch and smell to navigate and hunt. This reliance on tactile and olfactory senses makes the star-nosed mole highly specialized for its niche habitat.

The unique characteristics of the star-nosed mole have made it a subject of interest in various scientific fields. Neuroscientists study its extraordinary touch organ to understand how the brain processes sensory information. The mole's rapid prey identification and consumption have also intrigued researchers studying feeding behaviors and adaptations in mammals.

Despite its specialized adaptations, the star-nosed mole faces several challenges in its natural habitat. Wetland drainage, pollution, and habitat destruction pose significant threats to its population. Conservation efforts aimed at protecting wetland environments are crucial for the survival of this and other wetland-dependent species. Preserving these habitats ensures the continued existence of the star-nosed mole and the rich biodiversity of wetland ecosystems.

In popular culture, the star-nosed mole often captures the imagination due to its bizarre appearance. It has been featured in documentaries, educational programs, and nature photography, highlighting the diversity and wonder of the natural world. The mole's distinctive nose serves as a reminder of the incredible variety of adaptations that have evolved in response to specific environmental pressures.

In conclusion, the star-nosed mole stands out as one of the strangest and most specialized creatures in the animal kingdom. Its unique star-shaped nose, rapid prey detection, and adaptations to wetland habitats make it a remarkable example of evolutionary innovation. The study and conservation of the star-nosed mole contribute to our understanding of sensory biology and the importance of protecting diverse ecosystems.

7. Tarsier

The tarsier is a small primate that inhabits the forests of Southeast Asia, including the Philippines, Borneo, and Sumatra. Known for its enormous eyes and long fingers, the tarsier is one of the most distinctive and peculiar mammals in the world. These nocturnal creatures have evolved several unique adaptations that enable them to thrive in their arboreal habitats.

Tarsiers have the largest eyes relative to their body size of any mammal. Each eye is approximately the same size as the tarsier's brain, allowing them to see exceptionally well in low-light conditions. This adaptation is crucial for their nocturnal lifestyle, as they rely heavily on their keen vision to hunt insects and small vertebrates at night. Unlike other nocturnal animals that use a reflective layer behind their retinas to enhance night vision, tarsiers have a high density of rod cells in their eyes, giving them excellent visual acuity.

In addition to their remarkable eyes, tarsiers possess elongated fingers and toes tipped with adhesive pads, enabling them to grip branches and navigate the forest canopy with ease. Their long, slender limbs and powerful hind legs allow them to leap great distances from tree to tree, an essential skill for evading predators and capturing prey. Tarsiers are capable of rotating their heads almost 180 degrees, much like owls, which helps them scan their surroundings for potential threats and opportunities.

The diet of tarsiers primarily consists of insects, but they also prey on small birds, bats, and reptiles. Their hunting technique involves silently stalking their prey before making a rapid, precise leap to capture it with their sharp teeth. This method of hunting requires a combination of stealth, agility, and quick reflexes, making tarsiers highly efficient predators.

Tarsiers are also known for their vocalizations, which play a crucial role in communication and social interactions. They produce a variety of sounds, including high-pitched calls and ultrasonic vocalizations that are inaudible to humans. These vocalizations are used to establish territory, attract mates, and coordinate group movements. Tarsiers are generally solitary animals, but some species form small family groups, particularly during the breeding season.

The reproductive biology of tarsiers is equally fascinating. They typically give birth to a single offspring after a gestation period of about six months. Newborn tarsiers are relatively well-developed, with open eyes and the ability to cling to their mother's fur. The mother carries the infant in her mouth or on her belly while foraging, ensuring its safety and nourishment. Juvenile tarsiers grow rapidly and reach maturity within a year, ready to establish their own territories and continue the cycle of life.

Despite their unique adaptations and ecological importance, tarsiers face several conservation challenges. Habitat destruction due to deforestation, agriculture, and urbanization poses a significant threat to their populations. The illegal pet trade also endangers tarsiers, as their small size and cute appearance make them targets for capture and sale. Conservation efforts are essential to protect these extraordinary primates and their natural habitats.

Research on tarsiers provides valuable insights into primate evolution and the diversity of life in Southeast Asia's forests. By studying their behavior, physiology, and ecology, scientists can better understand the adaptive strategies that have enabled tarsiers to survive and thrive in their environment. Conservation programs aimed at preserving forest habitats and combating illegal wildlife trade are crucial for ensuring the long-term survival of tarsiers.

In conclusion, the tarsier is a remarkable example of evolutionary adaptation, with its enormous eyes, elongated limbs, and specialized hunting techniques. These small primates are a testament to the diversity and complexity of life on Earth, highlighting the importance of conserving the unique ecosystems they inhabit. The continued study and protection of tarsiers are vital for preserving the rich biodiversity of our planet.

8. Aye-Aye

The aye-aye (Daubentonia madagascariensis) is one of the most unusual and distinctive primates in the world, native to the island of Madagascar. This nocturnal lemur is instantly recognizable by its large eyes, bushy tail, and, most notably, its elongated middle finger. The aye-aye's unique adaptations have made it a subject of fascination and study, but also of local superstition and fear.

The aye-aye's elongated middle finger is perhaps its most remarkable feature. This thin, skeletal finger is highly flexible and is used primarily for foraging. The aye-aye employs a unique method called percussive foraging to locate insect larvae hidden within tree bark. It taps on the wood with its middle finger, listening for the hollow sounds that indicate the presence of cavities. Once it detects a cavity, the aye-aye uses its incisors to gnaw a hole in the wood, and then inserts its long finger to extract the larvae. This specialized foraging technique is an adaptation that allows the aye-aye to exploit food sources that are inaccessible to most other animals.

The aye-aye's large, sensitive ears play a crucial role in its foraging behavior. These ears can rotate independently, allowing the aye-aye to precisely locate the sounds of its prey within the trees. This remarkable hearing ability, combined with its keen sense of touch, makes the aye-aye an extraordinarily effective nocturnal hunter.

In addition to its foraging adaptations, the aye-aye has other distinctive features. Its teeth are constantly growing, similar to those of rodents, which helps it maintain the ability to gnaw through tough materials like wood and bamboo. The aye-aye's large eyes provide excellent night vision, essential for its nocturnal lifestyle. Its bushy tail, longer than its body, aids in balance as it moves through the forest canopy.

The aye-aye's diet is varied and includes not only insect larvae but also fruits, nuts, seeds, and nectar. It has a particular fondness for

25

coconuts and can gnaw through the tough husks to access the nutritious flesh inside. This diverse diet helps the aye-aye to survive in different habitats across Madagascar, from rainforests to dry deciduous forests.

Despite its fascinating adaptations, the aye-aye is often regarded with fear and superstition by local communities in Madagascar. In some cultures, it is believed that seeing an aye-aye is an omen of death, leading to the animal being killed on sight. This cultural stigma, combined with habitat destruction due to deforestation, poses a significant threat to the aye-aye's survival. Conservation efforts are crucial to protect this unique species and its habitat, promoting coexistence and understanding between local communities and wildlife.

The aye-aye is currently classified as Endangered by the International Union for Conservation of Nature (IUCN). Conservation programs aim to protect the aye-aye by preserving its natural habitat, raising awareness about its ecological importance, and combating the myths and superstitions that lead to its persecution. Captive breeding programs and research initiatives also play a role in ensuring the survival of this extraordinary primate.

Research on the aye-aye provides valuable insights into primate evolution and the ecological dynamics of Madagascar's forests. By studying its behavior, physiology, and ecology, scientists can better understand the adaptive strategies that have enabled the aye-aye to thrive in its unique environment. The aye-aye's role as a seed disperser and insect predator also highlights its importance in maintaining the health and balance of Madagascar's ecosystems.

In popular culture, the aye-aye often captures the imagination due to its bizarre appearance and intriguing behaviors. Documentaries, books, and educational programs frequently feature the aye-aye, helping to raise awareness about its conservation needs and the rich biodiversity of Madagascar.

In conclusion, the aye-aye stands out as one of the strangest and most specialized primates in the world. Its elongated middle finger, percussive foraging technique, and unique adaptations make it a remarkable example of evolutionary innovation. The conservation of the aye-aye is crucial for preserving the biodiversity of Madagascar and understanding the complex interactions within its ecosystems.

9. Goblin Shark

The goblin shark (Mitsukurina owstoni) is one of the most bizarre and enigmatic creatures in the ocean. Often referred to as a "living fossil," the goblin shark has a lineage that dates back around 125 million years. This deep-sea shark is distinguished by its elongated, flattened snout, protruding jaws, and pinkish-gray skin, giving it a ghastly and otherworldly appearance.

The goblin shark's most notable feature is its long, blade-like snout, which is covered in specialized sensory organs called ampullae of Lorenzini. These organs detect the electric fields produced by the movements of prey, making the goblin shark an effective hunter in the dark depths of the ocean. The shark's jaws are highly protrusible, meaning they can extend outward dramatically to catch prey. This jaw structure allows the goblin shark to quickly snap up fish, squid, and crustaceans that venture too close.

Goblin sharks inhabit deep waters, typically found at depths ranging from 200 to 1,300 meters (656 to 4,265 feet). They are most commonly seen in the waters off Japan, where they were first discovered, but their range extends to various parts of the Atlantic, Pacific, and Indian Oceans. Due to the extreme depths at which they live, goblin sharks are rarely encountered by humans and much of their behavior and biology remains a mystery.

The coloration of the goblin shark is another intriguing aspect of its appearance. Its pinkish hue is due to the transparency of its skin and the visible blood vessels beneath it. This unique coloration provides camouflage in the dim, blue light of the deep sea, helping the shark to blend in with its surroundings and avoid detection by both predators and prey.

The goblin shark's slow, sluggish movements are well-suited to its deep-sea environment, where conserving energy is crucial. Unlike their more active, fast-swimming relatives, goblin sharks rely on ambush tactics to capture prey. They remain motionless, waiting for unsuspecting creatures to come within range before rapidly extending their jaws to seize them. This hunting strategy is effective in the low-energy, food-scarce environments of the deep ocean.

Despite its fearsome appearance, the goblin shark poses little threat to humans due to its deep-sea habitat and reclusive nature. However, it occasionally gets caught as bycatch in deep-sea fishing operations. The rarity of goblin shark sightings and specimens makes each encounter valuable for scientific research, providing insights into the biology and ecology of this ancient species.

Research on goblin sharks has revealed some fascinating aspects of their physiology and evolution. Their unique jaw mechanism, for instance, offers clues about the evolutionary adaptations that have allowed them to survive for millions of years. Studies of their sensory systems contribute to our understanding of how deep-sea creatures navigate and hunt in environments with little to no light. The goblin shark's status as a "living fossil" highlights the incredible diversity and resilience of life in the deep ocean. It serves as a reminder of the ancient origins of many marine species and the evolutionary processes that continue to shape life in the ocean's depths. Protecting deep-sea habitats is essential for preserving these ancient lineages and the unique biodiversity they represent.

In popular culture, the goblin shark has captured the imagination of many due to its grotesque appearance and mysterious nature. It has been featured in documentaries, books, and even horror films, often depicted as a monstrous, alien-like creature. While its appearance may be unsettling, the goblin shark is a fascinating example of the extraordinary adaptations that life can develop in response to extreme environments.

In conclusion, the goblin shark is one of the strangest and most intriguing creatures in the ocean. Its unique adaptations, ancient lineage, and elusive nature make it a subject of fascination and study. The conservation of deep-sea environments is crucial for ensuring the survival of goblin sharks and other remarkable species that dwell in the depths of our oceans.

10. Glaucus Atlanticus

The Glaucus atlanticus, commonly known as the blue dragon or blue sea slug, is one of the ocean's most strikingly beautiful and unusual creatures. This small, blue mollusk is a type of nudibranch, a group of soft-bodied, marine gastropod mollusks. The blue dragon is found floating on the surface of warm ocean waters around the world, particularly in the Atlantic, Pacific, and Indian Oceans.

The blue dragon's appearance is nothing short of spectacular. Its elongated, flattened body is adorned with finger-like appendages that radiate from its sides, resembling wings or fins. These cerata, as they are called, are arranged in such a way that they give the blue dragon a unique and elegant silhouette. The creature's vibrant blue and silver coloration helps it blend with the ocean's surface, providing camouflage from both aerial and aquatic predators.

Glaucus atlanticus is a pelagic species, meaning it spends its life drifting on the ocean's surface. It achieves buoyancy by swallowing air and storing it in its stomach, allowing it to float upside down.

This floating lifestyle makes the blue dragon a formidable predator in its own right, despite its small size of about 3 centimeters (1.2 inches) in length.

One of the most fascinating aspects of the blue dragon is its diet. It feeds on venomous siphonophores, such as the Portuguese man o' war, as well as other pelagic organisms. The blue dragon has developed a remarkable adaptation that allows it to consume these venomous prey items without harm. It selectively stores the venomous nematocysts (stinging cells) from its prey in specialized sacs at the tips of its cerata. When threatened, the blue dragon can deploy these nematocysts to deliver a painful sting to potential predators, effectively using its prey's venom as a defense mechanism.

This ability to incorporate venom into its own defense system makes the blue dragon a unique and formidable creature in the marine world. Its sting can be painful to humans, although encounters are relatively rare due to its open ocean habitat. However, beachgoers should exercise caution if they come across a blue dragon washed ashore, as the venom retained from its prey can still cause a painful reaction.

The life cycle of the blue dragon is also intriguing. They are hermaphroditic, meaning each individual possesses both male and female reproductive organs. When two blue dragons' mate, they exchange sperm, and both individuals can lay eggs. The egg strings are attached to floating objects in the ocean, where they hatch into free-swimming larvae. These larvae eventually develop into the adult blue dragons, continuing the cycle of life in the open ocean.

Despite its relatively small size, the blue dragon plays an important role in the marine ecosystem. By preying on siphonophores and other pelagic organisms, it helps to control the populations of these species, maintaining the balance within its habitat. The blue dragon's presence also highlights the incredible diversity and complexity of life in the open ocean.

In recent years, the blue dragon has gained popularity and attention due to its striking appearance and unique adaptations. It has been featured in various marine biology studies, documentaries, and even social media posts, capturing the fascination of people around the world. This increased awareness can contribute to marine conservation efforts by highlighting the importance of preserving the delicate ecosystems that support such extraordinary creatures.

In conclusion, the Glaucus atlanticus, or blue dragon, is one of the most unusual and beautiful creatures in the ocean. Its vivid coloration, unique adaptations, and remarkable lifestyle make it a subject of fascination and study. The conservation of marine environments is crucial to ensure the survival of the blue dragon and the myriad other species that share its oceanic habitat.

Chapter 2: Bizarre Places

1.   The Bermuda Triangle

The Bermuda Triangle, also known as the Devil's Triangle, is a loosely defined region in the western part of the North Atlantic

Ocean. It is bounded by points in Miami, Bermuda, and Puerto Rico. This area has gained worldwide notoriety due to an unusual number of ships and aircraft that have mysteriously disappeared under unexplained circumstances. Over the years, the Bermuda Triangle has become synonymous with mystery, intrigue, and speculation.

The legend of the Bermuda Triangle began in the mid-20th century when reports of strange disappearances started to capture public attention. One of the earliest incidents occurred in 1918 when the USS Cyclops, a massive Navy cargo ship, vanished without a trace along with its 309 crew members. This incident was followed by numerous other disappearances, including the infamous Flight 19, a group of five torpedo bombers that went missing during a training mission in 1945. Despite extensive search and rescue operations, no wreckage or bodies were ever found, fueling speculation and theories about the region's mysterious powers.

Various explanations have been proposed to account for the unusual number of disappearances in the Bermuda Triangle. Some attribute the phenomenon to natural causes such as violent weather patterns, underwater earthquakes, and powerful ocean currents. The Gulf Stream, a major ocean current that flows through the Triangle, can cause rapid changes in weather and sea conditions, potentially leading to sudden and catastrophic accidents.

Another natural explanation involves methane hydrate deposits on the ocean floor. These deposits can produce massive bubbles of methane gas that, if released, could reduce the water's density and cause ships to sink rapidly. This theory suggests that such gas eruptions could also affect aircraft by creating turbulent air pockets.

In addition to natural causes, numerous supernatural and extraterrestrial theories have been proposed. Some believe that the Bermuda Triangle is a site of alien abductions or a gateway to another dimension. Others suggest that it is the location of the lost city of Atlantis, with advanced technology causing the disappearances. These more fantastical explanations have been popularized by books, movies, and television shows, further cementing the Bermuda Triangle's status as a modern myth.

Despite the many theories, there is no conclusive evidence to support any particular explanation for the disappearances in the Bermuda Triangle. The U.S. Navy and Coast Guard maintain that there is no greater frequency of incidents in the Triangle than in any other heavily traveled region of the world. They attribute most disappearances to human error, mechanical failure, and natural environmental factors.

Scientific investigations have also cast doubt on the Bermuda Triangle's reputation as a hotspot for mysterious occurrences. Studies have shown that the number of incidents in the Triangle is proportional to the amount of traffic it receives. Additionally, modern navigation technology and improved safety measures have significantly reduced the number of unexplained disappearances in recent years.

The enduring mystery of the Bermuda Triangle continues to captivate the public's imagination. Its combination of real-life incidents, speculative theories, and popular culture references creates a compelling narrative that appeals to our fascination with the unknown. Whether viewed as a place of supernatural activity or simply a region with challenging navigational conditions, the Bermuda Triangle remains one of the most enigmatic places on Earth.

In conclusion, the Bermuda Triangle is a region steeped in mystery and speculation. While natural explanations such as weather patterns and ocean currents provide plausible reasons for many of the disappearances, the lack of concrete evidence has allowed more fantastical theories to flourish. The Bermuda Triangle serves as a reminder of the power of the unknown to capture our imaginations and the enduring allure of unsolved mysteries.

2. The Door to Hell (Darvaza Gas Crater)

The Door to Hell, also known as the Darvaza Gas Crater, is a natural gas field in the Karakum Desert of Turkmenistan that has been burning continuously since 1971. This eerie and captivating site, with its perpetual flames and glowing crater, has earned the nickname "The Door to Hell" due to its otherworldly appearance and the infernal impression it leaves on visitors.

The origins of the Darvaza Gas Crater can be traced back to a Soviet drilling operation in 1971. Geologists were drilling in the area when they inadvertently punctured a large underground natural gas cavern. The ground beneath the drilling rig collapsed, forming a massive crater approximately 70 meters (230 feet) in diameter and 20 meters (66 feet) deep. To prevent the release of dangerous methane gas into the atmosphere, the geologists decided to ignite the gas, expecting it to burn off within a few weeks. However, the fire has been burning ever since, creating a continuous inferno that has lasted for decades.

The sight of the Darvaza Gas Crater at night is both mesmerizing and surreal. The flames can be seen from miles away, casting an eerie glow across the desert landscape. The intense heat and the constant roar of the fire create an atmosphere that feels almost supernatural. Visitors who brave the journey to this remote location are often struck by the stark contrast between the barren desert and the blazing crater, which appears like a portal to another world.

The Darvaza Gas Crater has become a popular tourist attraction, drawing adventurous travelers from around the world who are eager to witness this geological marvel. Despite its remote location and the challenging conditions of the Karakum Desert, the crater's

34

fiery spectacle continues to captivate those who make the trek. The site has also been featured in numerous documentaries and travel shows, further cementing its status as one of the world's most unusual and fascinating destinations.

In addition to its visual appeal, the Darvaza Gas Crater holds scientific significance. It serves as a stark reminder of the potential hazards associated with drilling for natural resources and the unforeseen consequences that can arise from such activities. The crater also provides a unique opportunity for researchers to study the long-term effects of natural gas combustion and the environmental impact of such phenomena.

Efforts to extinguish the flames have been considered over the years, but no definitive plans have been implemented. The Turkmenistan government has expressed interest in closing the crater to promote natural gas production in the region, but the logistical challenges and the potential risks involved have delayed any action. For now, the Darvaza Gas Crater continues to burn, serving as a symbol of both human intervention and the untamed power of nature.

In recent years, the Turkmenistan government has sought to capitalize on the crater's popularity by promoting it as a unique tourist destination. Infrastructure improvements, such as the construction of a nearby campground and visitor facilities, aim to make the site more accessible and comfortable for tourists. These efforts are part of a broader strategy to boost tourism and showcase Turkmenistan's natural wonders to the world.

The story of the Darvaza Gas Crater, or the Door to Hell, is a testament to the unpredictable and often dramatic interactions between human activities and the natural environment. It highlights the need for caution and respect when exploiting natural resources and underscores the importance of understanding the potential long-term impacts of such actions.

In conclusion, the Door to Hell is one of the most bizarre and captivating places on Earth. Its perpetual flames, eerie glow, and remote desert location create a hauntingly beautiful spectacle that continues to intrigue and inspire awe. As a natural wonder and a reminder of the consequences of human intervention, the Darvaza Gas Crater stands as a unique and enduring symbol of the extraordinary forces that shape our world.

3.   Pamukkale, Turkey

Pamukkale, located in southwestern Turkey, is one of the world's most surreal and beautiful natural wonders. Known as the "Cotton Castle" in Turkish, Pamukkale is famous for its stunning white terraces and thermal pools, which create a breathtaking landscape that looks like a frozen waterfall. This unique site has been a popular attraction for thousands of years, drawing visitors from all over the world to bathe in its mineral-rich waters and marvel at its otherworldly beauty.

The terraces of Pamukkale are formed by the deposition of calcium carbonate, which is carried by the hot springs that flow down the hillside. Over time, the calcium carbonate solidifies into travertine, creating the cascading terraces and pools that characterize the site. The thermal waters of Pamukkale have been valued for their therapeutic properties since ancient times, with people traveling from far and wide to bathe in the warm, mineral-rich waters, believed to cure various ailments and promote overall health.

Pamukkale's natural beauty is complemented by its historical significance. The ancient Greco-Roman city of Hierapolis, built on the plateau above the terraces, adds an extra layer of cultural and historical interest to the site. Hierapolis was founded in the 2nd century BCE and quickly became a prominent spa city, attracting visitors seeking the healing benefits of its thermal waters. The city thrived for centuries, with impressive structures such as the large

Roman theater, the Temple of Apollo, and the Necropolis, which is one of the best-preserved ancient cemeteries in Turkey.

The combination of Pamukkale's natural and historical wonders led to its designation as a UNESCO World Heritage Site in 1988. Efforts have been made to preserve both the natural terraces and the ruins of Hierapolis, ensuring that future generations can continue to enjoy and learn from this extraordinary place.

Visiting Pamukkale is a unique experience, offering the opportunity to walk barefoot along the terraces, feeling the warm water flow around your feet, and to relax in the thermal pools. The terraces are a popular spot for photography, with their dazzling white formations and turquoise waters providing a striking contrast against the surrounding landscape. In addition to exploring the terraces, visitors can also tour the ancient ruins of Hierapolis, gaining insight into the rich history of this once-thriving spa city.

Pamukkale's thermal waters are still enjoyed by visitors today, with several modern facilities offering thermal baths and spa treatments. The nearby town of Pamukkale provides accommodation and amenities for tourists, making it a convenient base for exploring the area. The blend of natural beauty, historical intrigue, and modern comfort makes Pamukkale a destination that appeals to a wide range of travelers.

The preservation of Pamukkale is crucial for maintaining its status as a natural and cultural treasure. Measures have been implemented to protect the terraces from damage caused by tourism, including restricting access to certain areas and regulating the flow of water to prevent erosion. These efforts are essential for ensuring that Pamukkale remains a pristine and captivating site for future visitors.

In conclusion, Pamukkale, with its stunning white terraces and rich historical heritage, is one of the most beautiful and unusual places on Earth. Its unique combination of natural wonder and ancient history creates a magical experience that leaves a lasting impression on all who visit. The conservation of Pamukkale is vital for preserving its beauty and significance, allowing it to continue to inspire and enchant people from around the world.

4. Lake Hillier, Australia

Lake Hillier, located on Middle Island off the coast of Western Australia, is one of the most striking natural wonders in the world due to its vivid pink color. This saline lake, surrounded by lush eucalyptus and paperback trees, creates a stunning contrast against the deep blue of the surrounding ocean, making it a truly unique and picturesque sight.

The vibrant pink color of Lake Hillier is its most defining feature, and the exact cause of this phenomenon has intrigued scientists and visitors alike. The prevailing theory is that the color is due to the presence of Dunaliella salina, a type of microalgae that produces carotenoids, including beta-carotene, which has a reddish-pink hue. Additionally, the lake contains high concentrations of halophilic bacteria, which thrive in the saline environment and also contribute to the lake's distinctive color.

Unlike other pink lakes around the world, such as the Pink Lake in Senegal or the Laguna Colorada in Bolivia, Lake Hillier retains its pink color year-round and does not change significantly with variations in temperature or sunlight. The consistency of its color, combined with the surrounding lush vegetation and white salt crust along the shoreline, creates a surreal and captivating landscape.

Lake Hillier was first discovered in 1802 by British explorer Matthew Flinders during his expedition to map the coastline of Australia. He noted the lake's unusual color in his journal, describing it as a "rose-colored" body of water. Since then, Lake Hillier has fascinated both explorers and scientists, becoming a popular subject of study and tourism.

The lake is relatively small, measuring about 600 meters in length and 250 meters in width. It is surrounded by dense vegetation, including a thick forest of eucalyptus and paperback trees, which adds to the beauty and mystery of the site. The remoteness of Lake Hillier has helped preserve its pristine condition, as it is accessible only by boat or air.

Visitors to Lake Hillier often take scenic flights over the lake to capture aerial views of its vibrant pink waters contrasting with the blue ocean and green forest. Boat tours are also available, providing an opportunity to see the lake up close and experience its tranquil beauty. Swimming in Lake Hillier is generally discouraged to preserve its natural state and because the high salt content can be harsh on the skin.

The scientific study of Lake Hillier has provided valuable insights into the extremophiles – organisms that thrive in extreme conditions – that inhabit the lake. The unique ecosystem of Lake Hillier serves as a natural laboratory for researchers studying the adaptation and survival strategies of halophilic microorganisms. Understanding these organisms can have broader implications for fields such as astrobiology, where scientists investigate the potential for life in extreme environments on other planets.

Conservation efforts are essential to protect Lake Hillier and its unique ecosystem. The lake and the surrounding area are part of the Recherche Archipelago Nature Reserve, which is managed by the Western Australian Department of Biodiversity, Conservation and Attractions. This designation helps ensure that the natural beauty and ecological significance of Lake Hillier are preserved for future generations.

In recent years, Lake Hillier has gained popularity through social media and travel blogs, attracting a growing number of visitors eager to witness its remarkable pink waters. While tourism brings attention and appreciation to this natural wonder, it also underscores the need for responsible tourism practices to minimize environmental impact and preserve the lake's pristine condition.

In conclusion, Lake Hillier stands out as one of the most bizarre and beautiful places on Earth. Its consistent pink color, coupled with the surrounding lush vegetation and striking contrasts, creates a landscape that is both surreal and enchanting. The scientific and ecological significance of Lake Hillier further enhances its appeal, making it a destination that continues to captivate and inspire awe. Preserving this unique natural wonder is crucial to ensure that its beauty and mystery endure for generations to come.

5.   The Catacombs of Paris, France

The Catacombs of Paris, also known as the "Empire of the Dead," are a vast network of underground tunnels and chambers beneath the streets of Paris, France. These catacombs hold the remains of over six million people, making them one of the world's largest and most macabre ossuaries. The history, scale, and eerie atmosphere of the catacombs have made them a popular destination for tourists and a subject of fascination and mystery.

The origins of the Paris Catacombs date back to the late 18th century, when the city faced a severe public health crisis due to overflowing cemeteries. The situation became so dire that many of the burial grounds, particularly the Cemetery of the Innocents, posed significant health risks to the residents. In response, the French government decided to relocate the remains from these cemeteries to the extensive network of underground quarries that had existed since Roman times.

The transfer of bones began in 1786 and continued until 1788. Workers moved the remains of millions of Parisians from various cemeteries, creating an ossuary in the former quarries. The bones were carefully arranged in walls and patterns, creating a haunting and morbidly beautiful spectacle. Skulls and femurs were often

40

used to create decorative structures, such as crosses and columns, adding to the catacombs' eerie charm.

The catacombs were officially opened to the public in the early 19th century, and they quickly became a popular attraction for those seeking to explore the dark underbelly of Paris. The entrance to the catacombs is located in the Montparnasse district, and visitors must descend a spiral staircase to reach the underground passages. The guided tour takes visitors through approximately 1.5 kilometers (1 mile) of the catacombs, showcasing some of the most famous and haunting sections of the ossuary.

The atmosphere within the catacombs is undeniably eerie. The narrow, dimly lit tunnels are lined with neatly stacked bones and skulls, creating an overwhelming sense of mortality and history. The air is cool and damp, and the silence is broken only by the echoes of footsteps and whispered voices. For many, the experience of walking through the catacombs is both humbling and chilling, a stark reminder of the city's ancient and tumultuous past. In addition to the public sections of the catacombs, there are miles of restricted tunnels and chambers that remain off-limits to visitors. These areas have become the subject of urban legends and speculation, with stories of secret societies, hidden treasures, and ghostly apparitions adding to the catacombs' mystique. Some adventurous explorers, known as "cataphiles," illegally venture into these forbidden areas, documenting their findings and experiences. The Catacombs of Paris are not only a historical and cultural landmark but also a site of ongoing scientific research. The bones within the catacombs provide valuable insights into the health, diet, and lifestyles of Parisians from different historical periods. Anthropologists and archaeologists study the remains to learn more about the city's population and how it evolved over time.

Despite their macabre nature, the catacombs are a significant part of Paris's heritage, attracting hundreds of thousands of visitors

41

each year. The preservation and maintenance of the catacombs are overseen by the City of Paris, which ensures that this unique historical site remains accessible and safe for future generations.

In conclusion, the Catacombs of Paris are one of the most bizarre and fascinating places in the world. Their vast network of tunnels, filled with the bones of millions, creates an eerie yet captivating experience for those who visit. The catacombs stand as a testament to Paris's rich history and the lengths to which the city has gone to address its challenges. As a place of both historical significance and eerie beauty, the catacombs continue to intrigue and inspire awe in all who venture into their depths.

6.   Salar de Uyuni, Bolivia

Salar de Uyuni, located in southwest Bolivia, is the world's largest salt flat, spanning over 10,000 square kilometers (3,900 square miles). This vast, otherworldly expanse is a place of surreal beauty and extreme contrasts, drawing tourists and photographers from around the globe. Formed from the prehistoric Lake Minchin, Salar de Uyuni is now a nearly flat, bright white landscape that creates striking reflections and optical illusions.

During the dry season, the salt flat is an endless, glistening expanse of hexagonal salt tiles, which are the result of the natural evaporation process. The intense whiteness of the salt crust is blinding under the sun, creating a stark, alien landscape that seems to stretch infinitely. This vast openness, coupled with the cloudless skies, offers unparalleled opportunities for stargazing, making Salar de Uyuni a dream destination for both astronomers and photographers.

In the rainy season, typically between January and March, the salt flat is transformed into a giant mirror. A thin layer of rainwater covers the flat, reflecting the sky and creating a stunning visual effect where the horizon disappears, and the earth and sky merge seamlessly. This mirror effect is one of the most iconic and photographed phenomena in Salar de Uyuni, offering a unique and mesmerizing experience for visitors.

Salar de Uyuni is also known for its unique geographical features and natural landmarks. One of the most notable is Isla Incahuasi, often referred to as Fish Island, an isolated outcrop in the middle of the salt flat. This island is covered with giant cacti, some of which are over 1,000 years old and reach heights of up to 12 meters (39 feet). The island provides a striking contrast to the flat, white expanse surrounding it and offers panoramic views of the salt flat from its highest points.

The salt flat is also home to a variety of wildlife, particularly during the wet season when it becomes a critical breeding ground for several species of flamingos. The sight of these vibrant pink birds against the stark white landscape adds another layer of surreal beauty to the area. Other species that inhabit the region include the Andean fox and various types of hummingbirds.

Salar de Uyuni is not just a place of natural wonder but also of significant economic importance. The salt flat contains vast quantities of lithium, an essential component in batteries for electronics and electric vehicles. Bolivia is estimated to hold over 7% of the world's known lithium reserves, and there are ongoing efforts to develop this resource sustainably. The extraction and processing of lithium must be managed carefully to minimize environmental impact and preserve the unique landscape of Salar de Uyuni.

Tourism is a major industry in the region, with guided tours taking visitors across the salt flat to explore its many wonders. These tours often include visits to the train cemetery near the town of Uyuni, where the remains of old steam locomotives rust in the desert, adding to the area's haunting beauty. Tours also typically feature overnight stays in salt hotels, unique accommodations built entirely from blocks of salt, providing a one-of-a-kind experience for travelers.

Efforts to protect and preserve Salar de Uyuni are crucial to maintain its natural beauty and ecological significance. Sustainable tourism practices and responsible lithium extraction are essential to ensure that this unique landscape remains pristine for future generations. Conservation initiatives aim to balance economic development with environmental stewardship, recognizing the importance of Salar de Uyuni as both a natural wonder and a valuable resource.

In conclusion, Salar de Uyuni is one of the most bizarre and breathtaking places on Earth. Its vast salt flats, mirror-like reflections, and unique geographical features create a landscape of surreal beauty and wonder. The salt flat's ecological, economic, and cultural significance makes it a vital part of Bolivia's heritage. Preserving Salar de Uyuni is essential to maintaining its status as one of the world's most extraordinary natural wonders, ensuring that its beauty and mystery continue to inspire awe for generations to come.

7. Mount Roraima, Venezuela

Mount Roraima, located in the Guiana Highlands of South America, is one of the most awe-inspiring and enigmatic natural formations in the world. Straddling the borders of Venezuela, Brazil, and Guyana, this tepui (tabletop mountain) stands as a majestic and isolated plateau rising dramatically above the surrounding rainforest. Its sheer cliffs, flat summit, and unique ecosystem make Mount Roraima a place of great scientific interest and natural beauty.

Mount Roraima is part of the larger Roraima Formation, a geological structure that dates back to the Precambrian era, over two billion years ago. The mountain's distinctive flat top and vertical cliffs, which reach heights of up to 400 meters (1,300 feet), give it an otherworldly appearance. The summit of Mount Roraima covers an area of approximately 31 square kilometers (12 square miles) and is often shrouded in clouds, adding to its mystique.

The journey to the summit of Mount Roraima is a challenging and rewarding adventure. The most popular route starts in the village of Paraitepui in Venezuela, where hikers trek through the Gran Sabana, a vast savanna dotted with unique rock formations and waterfalls. The ascent involves navigating steep trails, rocky terrain, and lush forests before reaching the base of the tepui. From there, climbers must tackle a demanding scramble up the cliffs to reach the summit. The entire trek typically takes five to seven days, requiring physical endurance and careful preparation.

Upon reaching the summit, visitors are rewarded with an unparalleled landscape of surreal beauty. The plateau is characterized by bizarre rock formations, deep crevices, and crystal-clear pools. These features have been sculpted over millions of years by erosion and weathering, creating a unique and alien environment. The summit is also home to a variety of endemic plant and animal species that have adapted to the harsh conditions. Carnivorous plants such as the pitcher plant and sundew thrive in the nutrient-poor soil, while unique species of frogs and lizards can be found in the isolated pools and crevices.

One of the most fascinating aspects of Mount Roraima is its role in indigenous mythology. The Pemon and Kapon tribes of the region regard the mountain as a sacred place, often referring to it as the "House of the Gods." According to their legends, Mount Roraima is the stump of a mighty tree that once bore all the fruits and vegetables of the world. The tree was felled by the trickster god Makunaima, causing a great flood that shaped the landscape of the region.

Mount Roraima also holds a special place in popular culture and literature. It served as the inspiration for Sir Arthur Conan Doyle's 1912 novel "The Lost World," in which an expedition discovers prehistoric creatures living on a remote plateau. The mountain's unique and isolated ecosystem, combined with its striking appearance, has captured the imagination of adventurers, writers, and filmmakers alike.

The conservation of Mount Roraima is crucial for preserving its unique biodiversity and cultural heritage. The mountain and its surrounding areas are protected within the boundaries of Canaima National Park in Venezuela, a UNESCO World Heritage Site. Efforts to promote sustainable tourism and minimize environmental impact are essential to ensure that this natural wonder remains unspoiled for future generations.

In recent years, Mount Roraima has become an increasingly popular destination for eco-tourism and adventure travel. Guided tours and treks provide visitors with the opportunity to explore the mountain's unique landscape while learning about its geological, ecological, and cultural significance. The influx of tourism has brought economic benefits to the local communities, highlighting the importance of balancing conservation and sustainable development.

In conclusion, Mount Roraima is one of the most bizarre and captivating places on Earth. Its sheer cliffs, flat summit, and unique ecosystem create a landscape of unparalleled beauty and mystery. The mountain's cultural significance and role in indigenous mythology add to its allure, making it a destination that continues to inspire awe and wonder. Preserving Mount Roraima is essential to maintain its natural and cultural heritage, ensuring that its beauty and mystery endure for generations to come.

8.   The Nazca Lines, Peru

The Nazca Lines, located in the Nazca Desert of southern Peru, are one of the world's most enigmatic and fascinating archaeological mysteries. These ancient geoglyphs, created by the Nazca culture between 500 BCE and 500 CE, cover an area of nearly 1,000 square kilometers (385 square miles). The Nazca Lines consist of hundreds of individual figures, ranging from simple lines and geometric shapes to complex depictions of animals, plants, and human forms.

The creation of the Nazca Lines is a remarkable feat of engineering and artistic expression. The geoglyphs were made by removing the dark, oxidized stones on the desert surface to reveal the lighter-colored soil beneath. This contrast between the dark and light surfaces creates the distinctive lines and shapes that are visible from the air. The dry, windless climate of the Nazca Desert has helped preserve these lines for over a millennium, allowing modern observers to marvel at their scale and precision.

The most famous and recognizable figures among the Nazca Lines include the hummingbird, the spider, the monkey, and the astronaut. These intricate designs, some of which span hundreds of meters, are best viewed from above, leading to speculation about their purpose and the methods used to create them. The figures are so large that they were not discovered in their entirety until the advent of modern aerial photography in the 20th century.

The purpose of the Nazca Lines remains one of archaeology's greatest mysteries. Various theories have been proposed to explain their function, ranging from astronomical calendars and irrigation systems to religious or ceremonial significance. One popular theory suggests that the lines were created as part of a complex system of astronomical alignments, used by the Nazca people to track celestial events and agricultural cycles. Some researchers believe that the geoglyphs may have served as pilgrimage routes or ritual pathways, with the figures representing deities or mythological beings.

In addition to their potential astronomical and religious functions, the Nazca Lines may also have been used for water-related rituals. The Nazca culture inhabited one of the driest regions on Earth, and the availability of water was critical for their survival. Some scholars suggest that the lines were created as offerings to the gods in hopes of securing water and fertility for their crops.

The mystery of the Nazca Lines has captured the imagination of both scholars and the general public. Over the years, various alternative theories and speculative ideas have emerged, including suggestions that the lines were created by extraterrestrial beings or that they serve as markers for ancient alien landing sites. While these ideas are generally considered fringe theories, they have contributed to the enduring fascination with the Nazca Lines.

The preservation of the Nazca Lines is a significant concern, as the increasing popularity of tourism and the expansion of modern infrastructure pose threats to their integrity. In recent years, efforts have been made to protect and conserve the geoglyphs, including the establishment of the Nazca Lines as a UNESCO World Heritage Site in 1994. Ongoing research and conservation initiatives aim to safeguard this unique cultural heritage for future generations.

Visitors to the Nazca Lines can explore the site through various means, including scenic flights that provide an aerial view of the geoglyphs. Local tour operators offer guided trips that combine flights with ground tours of nearby archaeological sites and museums, providing a comprehensive understanding of the Nazca culture and their remarkable achievements. The Maria Reiche Museum, named after the German mathematician and archaeologist who dedicated her life to studying the lines, offers valuable insights into the history and significance of the geoglyphs.

In conclusion, the Nazca Lines of Peru are one of the most bizarre and captivating archaeological mysteries in the world. Their immense scale, intricate designs, and enigmatic purpose continue to intrigue and inspire awe. The preservation and study of the Nazca Lines are crucial for unlocking the secrets of this ancient culture and ensuring that these remarkable geoglyphs endure as a testament to human ingenuity and creativity.

9. The Doorway to Heaven, Tianmen Mountain, China

Tianmen Mountain, located in the Hunan Province of China, is renowned for its breathtaking natural beauty and the awe-inspiring Tianmen Cave, often referred to as the "Doorway to Heaven." This natural arch, which sits at an elevation of approximately 1,519 meters (4,984 feet), is one of the highest naturally occurring caves in the world and has become a symbol of the mountain's mystical allure.

The journey to the summit of Tianmen Mountain and the Doorway to Heaven is an adventure in itself. Visitors can reach the mountain via the Tianmen Mountain Cableway, which is one of the longest and most spectacular cable car rides in the world. The cableway stretches over 7,455 meters (24,459 feet) and ascends 1,279 meters (4,196 feet), offering passengers panoramic views of the surrounding landscape. The ride provides a unique perspective of the lush forests, steep cliffs, and winding roads that characterize the area.

One of the most remarkable features of Tianmen Mountain is the Tianmen Cave, a massive natural arch that measures approximately 131.5 meters (431 feet) in height and 57 meters (187 feet) in width. The cave was formed through a series of natural geological processes over millions of years and has become a major attraction for tourists and pilgrims alike. To reach the cave, visitors must climb a steep staircase of 999 steps, symbolizing the ascent to heaven. The climb is physically demanding but offers a sense of achievement and awe upon reaching the top.

In addition to the Doorway to Heaven, Tianmen Mountain is home to several other attractions that highlight its natural and cultural significance. The mountain is dotted with ancient temples, such as the Tianmen Mountain Temple, which dates back to the Tang Dynasty (618-907 CE). This temple is a place of spiritual reflection and offers visitors a glimpse into the region's rich religious heritage. Another popular attraction is the Tianmen Mountain Glass Skywalk, a transparent walkway that clings to the edge of a cliff and offers breathtaking views of the valley below. Walking along the glass skywalk is an exhilarating experience, providing a unique perspective of the mountain's sheer drop-offs and dramatic landscapes. For the adventurous, the Coiling Dragon Cliff skywalk and the cliff-hanging walkways offer additional thrilling opportunities to explore the mountain's vertical terrain.

Tianmen Mountain is also famous for its winding mountain road, known as the "Heaven-Linking Avenue" or the "99 Bends Road." This serpentine road features 99 sharp turns and ascends nearly 11 kilometers (6.8 miles) from the base of the mountain to the top. Driving or taking a bus along this road is an unforgettable experience, with each turn revealing new and stunning views of the surrounding landscape.

The unique combination of natural beauty, cultural heritage, and thrilling attractions has made Tianmen Mountain a popular destination for tourists from around the world. The mountain's diverse ecosystem, which includes rare plant and animal species, further adds to its allure. Efforts to preserve and protect the natural environment of Tianmen Mountain are crucial for maintaining its ecological balance and ensuring that future generations can continue to enjoy its wonders.

In conclusion, Tianmen Mountain and the Doorway to Heaven are among the most bizarre and captivating places on Earth. The mountain's dramatic landscapes, ancient temples, and unique attractions create a destination that is both mystical and awe-inspiring. The preservation of Tianmen Mountain's natural and cultural heritage is essential to maintaining its status as a symbol of beauty and spiritual significance. As visitors ascend the steps to the Doorway to Heaven, they are reminded of the enduring allure of nature's wonders and the human spirit's quest for the divine.

10. The Doorway to Heaven, Tianmen Mountain, China

Tianmen Mountain, located in the Hunan Province of China, is renowned for its breathtaking natural beauty and the awe-inspiring Tianmen Cave, often referred to as the "Doorway to Heaven." This natural arch, which sits at an elevation of approximately 1,519 meters (4,984 feet), is one of the highest naturally occurring caves in the world and has become a symbol of the mountain's mystical allure.

The journey to the summit of Tianmen Mountain and the Doorway to Heaven is an adventure in itself. Visitors can reach the mountain via the Tianmen Mountain Cableway, which is one of the longest and most spectacular cable car rides in the world. The cableway stretches over 7,455 meters (24,459 feet) and ascends 1,279 meters (4,196 feet), offering passengers panoramic views of the surrounding landscape. The ride provides a unique perspective of the lush forests, steep cliffs, and winding roads that characterize the area.

One of the most remarkable features of Tianmen Mountain is the Tianmen Cave, a massive natural arch that measures approximately

51

131.5 meters (431 feet) in height and 57 meters (187 feet) in width. The cave was formed through a series of natural geological processes over millions of years and has become a major attraction for tourists and pilgrims alike. To reach the cave, visitors must climb a steep staircase of 999 steps, symbolizing the ascent to heaven. The climb is physically demanding but offers a sense of achievement and awe upon reaching the top.

In addition to the Doorway to Heaven, Tianmen Mountain is home to several other attractions that highlight its natural and cultural significance. The mountain is dotted with ancient temples, such as the Tianmen Mountain Temple, which dates back to the Tang Dynasty (618-907 CE). This temple is a place of spiritual reflection and offers visitors a glimpse into the region's rich religious heritage.

Another popular attraction is the Tianmen Mountain Glass Skywalk, a transparent walkway that clings to the edge of a cliff and offers breathtaking views of the valley below. Walking along the glass skywalk is an exhilarating experience, providing a unique perspective of the mountain's sheer drop-offs and dramatic landscapes. For the adventurous, the Coiling Dragon Cliff skywalk and the cliff-hanging walkways offer additional thrilling opportunities to explore the mountain's vertical terrain.

Tianmen Mountain is also famous for its winding mountain road, known as the "Heaven-Linking Avenue" or the "99 Bends Road." This serpentine road features 99 sharp turns and ascends nearly 11 kilometers (6.8 miles) from the base of the mountain to the top. Driving or taking a bus along this road is an unforgettable experience, with each turn revealing new and stunning views of the surrounding landscape.

The unique combination of natural beauty, cultural heritage, and thrilling attractions has made Tianmen Mountain a popular destination for tourists from around the world. The mountain's diverse ecosystem, which includes rare plant and animal species, further adds to its allure. Efforts to preserve and protect the natural environment of Tianmen Mountain are crucial for maintaining its ecological balance and ensuring that future generations can continue to enjoy its wonders.

In conclusion, Tianmen Mountain and the Doorway to Heaven are among the most bizarre and captivating places on Earth. The mountain's dramatic landscapes, ancient temples, and unique attractions create a destination that is both mystical and awe-inspiring. The preservation of Tianmen Mountain's natural and cultural heritage is essential to maintaining its status as a symbol of beauty and spiritual significance. As visitors ascend the steps to the Doorway to Heaven, they are reminded of the enduring allure of nature's wonders and the human spirit's quest for the divine.

Chapter 3: Mystifying Animals

1.   The Axolotl (Ambystoma mexicanum)

The axolotl, also known as the Mexican walking fish, is one of the most intriguing and mystifying creatures in the animal kingdom. Despite its common name, the axolotl is not a fish but an amphibian, closely related to the tiger salamander. Native to the lake complex of Xochimilco in Mexico City, the axolotl is renowned for its unique appearance and extraordinary regenerative abilities.

Axolotls are neotenic, meaning they retain their larval features throughout their adult life. Unlike most amphibians, which undergo metamorphosis from larva to adult, axolotls remain aquatic and gilled. Their striking appearance includes a broad head, lidless eyes, and frilly external gills that protrude from either side of their head. These gills give the axolotl a distinctive and somewhat otherworldly look. They come in various colors, including wild-type (dark with greenish spots), leucistic (pale pink with red gills), albino, golden albino, and melanoid (all black).

Axolotls inhabit the remnants of ancient lakes and canals in Xochimilco, where they prefer still, deep waters with plenty of vegetation. Historically, these waters provided an ideal environment for axolotls, offering abundant food and protection from predators. However, the degradation of their natural habitat due to urbanization, pollution, and the introduction of invasive species has severely impacted their populations.

One of the most remarkable traits of the axolotl is its regenerative ability. Axolotls can regrow entire limbs, spinal cords, hearts, and other organs without any scarring. This extraordinary capability has made them a subject of significant interest in scientific research. Understanding the mechanisms behind their regenerative prowess could have profound implications for human medicine, particularly in the fields of tissue engineering and regenerative therapy. Researchers have identified certain genes and proteins that play crucial roles in the axolotl's regeneration process, offering insights that could one day lead to breakthroughs in treating human injuries and diseases.

In addition to their regenerative abilities, axolotls have an impressive capacity for healing and resistance to cancer. Their cells exhibit an unusual ability to avoid malignant transformation, which scientists are studying to develop potential cancer treatments. The axolotl's genome has been sequenced, revealing a wealth of information about the genetic basis of their unique traits and further enhancing their value as a model organism in scientific research.

Despite their fascinating biology, axolotls are critically endangered in the wild. Habitat loss, water pollution, and the introduction of predatory fish have all contributed to their decline. Conservation efforts are underway to protect the remaining wild populations, including habitat restoration projects and the creation of axolotl sanctuaries. Captive breeding programs have also been established to maintain genetic diversity and reintroduce axolotls into their natural habitats.

In addition to conservation initiatives, raising public awareness about the plight of the axolotl is crucial. Educational programs and collaborations with local communities aim to promote the importance of preserving this unique species and its habitat. By engaging local residents in conservation efforts, there is hope that the axolotl can be saved from extinction and continue to thrive in its natural environment.

The axolotl's role in Mexican culture and mythology also highlights its significance. Known as the "water monster" in the Nahuatl language, the axolotl is associated with the Aztec god Xolotl, who was believed to have transformed into an axolotl to escape death. This cultural heritage underscores the importance of preserving the axolotl not only for its biological uniqueness but also for its cultural and historical value.

In conclusion, the axolotl is one of the most mystifying and remarkable animals in the world. Its unique appearance, extraordinary regenerative abilities, and critical conservation status make it a species of immense scientific and cultural importance. Protecting the axolotl and its habitat is essential for preserving this incredible creature and unlocking the potential benefits of its remarkable biology for future generations.

2. The Platypus (Ornithorhynchus anatinus)

The platypus is one of the most extraordinary and unique animals on the planet. Native to the eastern regions of Australia, this small, semi-aquatic mammal defies classification with its unusual combination of features from different animal groups. The platypus, along with the echidna, belongs to the monotreme order, the only group of mammals that lay eggs instead of giving birth to live young.

At first glance, the platypus looks like a creature assembled from parts of various animals. It has a bill and webbed feet like a duck, a tail like a beaver, and a body covered in dense, waterproof fur similar to an otter. This bizarre appearance led early European naturalists to believe that the platypus was a hoax when it was first discovered in the late 18th century.

The platypus's duck-bill is one of its most distinctive features. Unlike a bird's beak, the platypus's bill is soft and covered with skin. It is highly sensitive and packed with electroreceptors, which allow the platypus to detect the electric fields generated by the muscle contractions of its prey. This adaptation is particularly useful for hunting underwater in murky conditions, where the platypus closes its eyes, ears, and nostrils and relies entirely on its bill to locate food.

Platypuses are carnivorous and primarily feed on aquatic invertebrates, such as insect larvae, worms, and crustaceans. They forage along the bottoms of rivers and streams, using their bill to sift through the substrate and detect prey. Once captured, the platypus stores the food in cheek pouches and returns to the surface to chew and swallow it.

One of the most surprising aspects of the platypus is that the males possess venomous spurs on their hind legs. During the breeding season, males use these spurs to deliver a potent venom to rivals or predators. While the venom is not lethal to humans, it can cause severe pain and swelling. This venomous trait is unique among mammals, further highlighting the platypus's unusual nature.

The platypus lays eggs, a characteristic that sets it apart from almost all other mammals. Females lay one to three eggs at a time and incubate them by curling around them in a burrow. The eggs hatch after about ten days, and the mother nurses the young by secreting milk through specialized mammary gland ducts, as platypuses lack nipples. The young platypuses, or puggles, remain in the burrow and continue to nurse for several months before venturing out on their own.

The platypus's combination of primitive and specialized traits makes it a key species for studying mammalian evolution. It

57

provides valuable insights into the early divergence of mammals from their reptilian ancestors and helps scientists understand the development of unique mammalian characteristics, such as lactation and endothermy.

Despite its fascinating biology, the platypus faces several conservation challenges. Habitat destruction, water pollution, and climate change threaten its populations. Urbanization and land clearing have reduced the availability of suitable habitats, while changes in water quality and flow can impact the availability of prey. Conservation efforts are focused on protecting and restoring habitats, monitoring populations, and mitigating threats to ensure the platypus's survival.

The platypus holds a special place in Australian culture and is an iconic symbol of the country's unique wildlife. It is featured on the Australian twenty-cent coin and is a popular subject in art, literature, and media. Public education and awareness campaigns aim to promote the conservation of the platypus and its habitats.

In conclusion, the platypus is one of the most mystifying and unique animals in the world. Its combination of characteristics from different animal groups, its venomous spurs, and its egg-laying habits make it a subject of endless fascination and scientific interest. Protecting the platypus and its habitats is essential for preserving this remarkable creature and continuing to learn from its evolutionary legacy.

3.   The Goblin Shark (Mitsukurina owstoni)

The goblin shark, often referred to as a "living fossil," is one of the most bizarre and elusive creatures in the ocean. This deep-sea shark, known for its distinctive appearance and ancient lineage, inhabits the dark depths of the world's oceans. Its unique physical characteristics and adaptations make it a fascinating subject of study and a source of endless intrigue.

The goblin shark's most striking feature is its long, flat snout, which extends forward like a blade. This protruding snout, called a rostrum, is covered in specialized sensory organs known as ampullae of Lorenzini. These organs allow the goblin shark to detect the electric fields produced by its prey, a crucial adaptation for hunting in the dark, deep-sea environment where light is scarce. When hunting, the goblin shark uses its rostrum to sense the presence of prey before rapidly extending its jaws to capture it.

The goblin shark's jaws are another remarkable adaptation. Unlike most sharks, the goblin shark has highly protrusible jaws, which can extend forward dramatically to snatch prey. This unique feeding mechanism is facilitated by elastic ligaments that allow the jaws to spring forward with great speed. The teeth of the goblin shark are long, slender, and needle-like, ideal for impaling soft-bodied prey such as fish, squid, and crustaceans.

Goblin sharks are typically pale pink or tan in color, a result of their translucent skin and visible blood vessels. Their coloration helps them blend into the dimly lit deep-sea environment. Adult goblin sharks can reach lengths of up to 3.8 meters (12.5 feet), although most specimens are smaller. Despite their fearsome appearance, goblin sharks are not considered a threat to humans due to their deep-sea habitat and elusive nature.

The goblin shark's habitat is primarily in the deep waters of continental slopes, submarine canyons, and seamounts, typically at depths ranging from 200 to 1,300 meters (656 to 4,265 feet). However, they have been recorded at depths as great as 1,370 meters (4,495 feet). The deep-sea environment is challenging to study, which has resulted in limited knowledge about the goblin shark's behavior, population size, and reproductive habits. Most of what is known about goblin sharks comes from specimens accidentally caught in deep-sea fishing nets or discovered by deep-sea submersibles and remote-operated vehicles (ROVs).

The goblin shark's lineage dates back to the Cretaceous period, around 125 million years ago, making it one of the oldest living shark species. Its primitive features and ancient origins have earned it the nickname "living fossil," and studying the goblin shark can provide valuable insights into the evolutionary history of sharks and other cartilaginous fish.

Despite its ancient lineage, the goblin shark is not immune to the threats faced by many deep-sea species. Bycatch in deep-sea fishing operations poses a significant risk to goblin sharks, as they can become entangled in trawling nets or longlines intended for other species. The deep-sea environment itself is also increasingly under threat from human activities, such as deep-sea mining and pollution, which can disrupt the delicate ecosystems that goblin sharks and other deep-sea creatures rely on.

Conservation efforts for deep-sea species like the goblin shark are challenging due to the difficulty of studying and monitoring these elusive animals. However, increased awareness of the importance of deep-sea ecosystems and the unique species they harbor is crucial for their protection. International cooperation and the implementation of sustainable fishing practices are essential steps toward ensuring the survival of the goblin shark and other deep-sea organisms.

In addition to conservation efforts, ongoing scientific research is vital for expanding our understanding of the goblin shark and its role in the deep-sea ecosystem. Advances in technology, such as deep-sea submersibles and genetic analysis, are helping scientists uncover more about this mysterious species. Each new discovery contributes to our knowledge of the goblin shark's biology, behavior, and evolutionary history.

In conclusion, the goblin shark is one of the most mystifying and unique animals in the ocean. Its distinctive appearance, ancient lineage, and remarkable adaptations make it a subject of great scientific interest and fascination. Protecting the goblin shark and its deep-sea habitat is essential for preserving this extraordinary species and the rich biodiversity of the ocean's depths. Continued research and conservation efforts are crucial for ensuring that the goblin shark remains a living testament to the wonders of the deep sea.

4.   The Saiga Antelope (Saiga tatarica)

The saiga antelope, with its distinctive humped nose and unique appearance, is one of the most peculiar and endangered mammals in the world. Native to the steppes of Central Asia, the saiga antelope is a symbol of the region's vast grasslands and has a rich history that dates back to the Pleistocene epoch. Despite its resilience, the saiga antelope faces numerous threats that have brought it to the brink of extinction.

The most striking feature of the saiga antelope is its large, bulbous nose, which hangs over its mouth like a short trunk. This unusual adaptation serves multiple purposes. During the summer, the nose helps filter out dust from the dry steppe environment, while in the winter, it warms the frigid air before it reaches the lungs. This specialized nose is crucial for the saiga's survival in the harsh and variable climate of the Central Asian steppes.

Saiga antelopes have a distinctive appearance, with a light sandy-brown coat that turns almost white in the winter. They have slender legs, a short tail, and long, pointed ears. Males are larger than females and possess lyre-shaped horns that are ridged and translucent. These horns are highly prized in traditional Chinese medicine, which has contributed to the saiga's decline due to illegal hunting.

Saigas are highly migratory animals, traveling vast distances across the steppes in search of food and suitable breeding grounds. They primarily feed on a variety of grasses, herbs, and shrubs, which

61

provide them with the necessary nutrients to survive in their challenging habitat. During migration, saigas can cover hundreds of kilometers, moving in large herds that offer protection from predators.

The saiga antelope's breeding season, known as the rut, occurs in late autumn. During this time, males compete for access to females, engaging in fierce battles using their horns. Dominant males gather harems of females and mate with them, ensuring the continuation of their genetic line. After a gestation period of about five months, females give birth to one or two calves, which are able to stand and follow their mother within a few hours of birth.

Despite their adaptability, saiga antelopes are critically endangered. Their populations have been decimated by a combination of factors, including habitat loss, poaching, disease, and climate change. The collapse of the Soviet Union in the early 1990s led to a dramatic increase in poaching, as economic instability and the breakdown of law enforcement made it easier for poachers to operate. Additionally, the demand for saiga horns in traditional medicine has fueled illegal hunting, further threatening the species. In recent years, disease outbreaks have had a catastrophic impact on saiga populations. In 2015, an outbreak of pasteurellosis, a bacterial infection, killed over 200,000 saigas in Kazakhstan, wiping out nearly two-thirds of the global population. Such outbreaks are exacerbated by climate change, which can alter the environment in ways that make diseases more likely to spread.

Conservation efforts are crucial to the survival of the saiga antelope. Various international and local organizations are working to protect the remaining populations through anti-poaching initiatives, habitat restoration, and disease management. Protected areas have been established in key saiga habitats, and breeding programs aim to increase population numbers and genetic diversity.

Public awareness and education campaigns are also essential for the conservation of the saiga antelope. By highlighting the importance of this unique species and the threats it faces, conservationists hope to garner support for protection efforts and reduce the demand for saiga horns in traditional medicine.

In conclusion, the saiga antelope is one of the most mystifying and endangered animals on the planet. Its distinctive appearance, migratory behavior, and resilience make it a fascinating species worthy of protection. The survival of the saiga antelope depends on continued conservation efforts, international cooperation, and public awareness. Preserving this remarkable species is essential for maintaining the biodiversity and ecological balance of the Central Asian steppes.

5.   The Mimic Octopus (Thaumoctopus mimicus)

The mimic octopus, a master of disguise and deception, is one of the most extraordinary creatures in the ocean. Discovered in the waters of Indonesia in 1998, this remarkable cephalopod has captivated scientists and divers alike with its ability to imitate a wide variety of marine species. The mimic octopus's unparalleled mimicry and adaptability make it a fascinating subject of study and a marvel of the natural world.

The mimic octopus is named for its incredible ability to mimic other sea creatures. Unlike other octopuses that primarily rely on camouflage, the mimic octopus can transform its shape, color, and behavior to imitate different species. This unique form of mimicry serves as both a defensive mechanism to avoid predators and a strategy to ambush prey.

One of the most remarkable aspects of the mimic octopus is the diversity of its impersonations. It can imitate a variety of marine animals, including lionfish, flatfish, sea snakes, jellyfish, and even crabs. When imitating a lionfish, for example, the mimic octopus

63

will spread its arms wide and fan them out to resemble the lionfish's venomous spines. To mimic a flatfish, it flattens its body and undulates along the seafloor. When impersonating a sea snake, it tucks most of its arms into a burrow and extends two arms in opposite directions, mimicking the movement and appearance of a venomous sea snake.

The mimic octopus's ability to change its appearance so drastically is due to specialized skin cells called chromatophores, iridophores, and leucophores. Chromatophores contain pigments that can be expanded or contracted to change color, while iridophores and leucophores reflect light to create iridescent and white effects. By controlling these cells, the mimic octopus can produce complex color patterns and textures that enhance its mimicry.

In addition to its physical transformations, the mimic octopus also alters its behavior to match the species it is imitating. This behavioral mimicry is crucial for convincing potential predators or prey that it is indeed another animal. For instance, when imitating a venomous sea snake, the mimic octopus will move its arms in a way that mimics the undulating motion of a sea snake, deterring potential predators.

The mimic octopus inhabits the shallow, muddy estuaries and coastal waters of Southeast Asia, where its mimicry skills are particularly useful. These environments are home to a wide variety of predators and prey, and the mimic octopus's ability to impersonate multiple species gives it a significant survival advantage. Its diet consists of small fish, crustaceans, and worms, which it captures using its dexterous arms and quick reflexes.

The discovery of the mimic octopus has provided valuable insights into the evolution of mimicry and camouflage in the animal kingdom. Scientists are particularly interested in understanding how the mimic octopus developed its extraordinary abilities and the genetic and neurological mechanisms underlying its mimicry.

Studying the mimic octopus can also shed light on the broader ecological interactions within its habitat and the evolutionary pressures that drive such complex behaviors.

Despite its remarkable abilities, the mimic octopus faces several threats in its natural habitat. Overfishing, habitat destruction, and pollution are significant concerns that impact the coastal ecosystems where the mimic octopus lives. Conservation efforts aimed at protecting these environments are essential for ensuring the survival of this extraordinary species.

In conclusion, the mimic octopus is one of the most mystifying and fascinating animals in the ocean. Its unparalleled ability to mimic a wide variety of marine species, combined with its adaptability and intelligence, makes it a subject of endless scientific interest and admiration. Protecting the mimic octopus and its habitat is crucial for preserving this remarkable example of nature's ingenuity and evolutionary creativity.

6. The Glass Frog (Centrolenidae)

The glass frog, a small and delicate amphibian native to the rainforests of Central and South America, is one of the most remarkable examples of nature's transparency. Named for its translucent skin, the glass frog offers a rare and fascinating glimpse into its inner workings. This unique trait, along with its arboreal lifestyle and distinctive reproductive behaviors, makes the glass frog a subject of great interest to scientists and nature enthusiasts alike.

Glass frogs belong to the family Centrolenidae, which includes more than 150 species. They are typically small, ranging from 2 to 8 centimeters (0.8 to 3.1 inches) in length. The most striking feature of these frogs is their translucent skin, particularly on the underside, through which their internal organs, including the heart,

65

liver, and gastrointestinal tract, can be seen. This transparency gives them a ghostly appearance and has led to their common name. The upper side of the glass frog is usually a vibrant green, which helps it blend into the lush foliage of its rainforest habitat. This camouflage is crucial for avoiding predators such as birds, snakes, and larger amphibians. When resting on leaves, the glass frog's green coloration makes it nearly invisible, while its transparent underside provides additional concealment.

The glass frog's reproductive behavior is as fascinating as its appearance. Mating typically occurs during the rainy season, when males call to attract females. The males produce a high-pitched, melodic call, often while perched on leaves or branches above streams. Once a female is attracted, the pair engages in amplexus, with the male clasping the female from behind. The female then lays her eggs on the underside of leaves overhanging water.

After laying the eggs, the female departs, leaving the male to guard the clutch. This parental care is relatively rare among amphibians and is crucial for the survival of the eggs. The male glass frog remains vigilant, protecting the eggs from predators such as insects and other amphibians. He also keeps the eggs moist by absorbing water through his skin and transferring it to the eggs, preventing them from drying out.

When the eggs hatch, the larvae drop into the water below, where they continue their development. The tadpoles of glass frogs are also somewhat transparent, although less so than the adults. They undergo metamorphosis in the water before emerging as fully formed frogs and beginning their arboreal life.

The unique transparency of glass frogs has intrigued scientists, leading to various studies on the potential evolutionary advantages and mechanisms behind this trait. One theory suggests that the transparency helps glass frogs avoid predation by breaking up their outline and making it more difficult for predators to detect them.

Another theory posits that the transparency could be a form of thermoregulation, allowing the frogs to absorb or dissipate heat more efficiently.

Despite their remarkable adaptations, glass frogs face several threats, primarily due to habitat loss and climate change. The destruction of rainforests for agriculture, logging, and urban development has significantly reduced their natural habitat. Additionally, climate change poses a threat by altering the delicate balance of their rainforest ecosystems, affecting the availability of water and the timing of breeding seasons.

Conservation efforts are essential to protect the glass frog and its habitat. These efforts include the establishment of protected areas, reforestation projects, and initiatives to reduce deforestation. Public awareness campaigns and education programs also play a vital role in promoting the conservation of these unique amphibians and their rainforest home.

In conclusion, the glass frog is one of the most mystifying and enchanting animals in the world. Its translucent skin, arboreal lifestyle, and dedicated parental care make it a fascinating subject of study and admiration. Protecting the glass frog and its rainforest habitat is crucial for preserving this extraordinary species and the rich biodiversity of its ecosystem.

7.   The Aye-Aye (Daubentonia madagascariensis)

The aye-aye, a nocturnal primate native to Madagascar, is one of the most unusual and enigmatic creatures in the animal kingdom. With its distinctive features and unique foraging behavior, the aye-aye has captivated scientists and intrigued the public. Despite its peculiar appearance, the aye-aye plays a crucial role in its ecosystem, and understanding its biology and conservation status is vital for preserving Madagascar's rich biodiversity.

The aye-aye is the world's largest nocturnal primate, with adults typically weighing between 2.5 to 2.8 kilograms (5.5 to 6.2 pounds). Its most striking feature is its elongated, thin middle finger, which it uses to tap on trees and locate insect larvae hidden within the wood. This specialized finger, combined with large, sensitive ears, enables the aye-aye to detect the hollow sounds of insect tunnels. Once it locates its prey, the aye-aye gnaws a hole in the wood with its rodent-like incisors and uses its elongated finger to extract the larvae.

The aye-aye's appearance is equally distinctive. It has a shaggy, dark brown or black coat, large eyes adapted for night vision, and a bushy tail that is longer than its body. Its hands and feet are equipped with sharp, curved claws that aid in climbing and foraging. This combination of traits gives the aye-aye a somewhat eerie appearance, contributing to its mystique and the superstitions surrounding it.

Historically, the aye-aye has been the subject of myths and legends in Madagascar. Many local communities view the aye-aye as an omen of bad luck or a harbinger of death. As a result, aye-ayes are often killed on sight, contributing to their decline. These superstitions, coupled with habitat loss due to deforestation, have placed the aye-aye at risk of extinction.

The aye-aye's diet primarily consists of insect larvae, but it also feeds on a variety of other foods, including fruits, nectar, seeds, and fungi. Its foraging behavior is highly specialized and involves a technique known as percussive foraging, where the aye-aye taps on tree bark to locate hollow cavities. This behavior is not only fascinating but also highlights the aye-aye's role as a key player in its ecosystem, helping to control insect populations and disperse seeds.

Aye-ayes are solitary animals, with individuals maintaining large territories that they mark with scent glands. They are highly arboreal, spending most of their lives in the treetops. During the day, they rest in spherical nests made of leaves and twigs, which they construct high in the canopy. These nests provide shelter and protection from predators such as the fossa, Madagascar's largest carnivorous mammal.

Reproduction in aye-ayes is not well understood, but it is known that females give birth to a single offspring after a gestation period of about 170 days. The young aye-aye is highly dependent on its mother for the first few months of life, gradually becoming more independent as it learns to forage and navigate the forest. The reproductive rate of aye-ayes is low, which, combined with their long lifespan of up to 23 years in the wild, makes population recovery slow.

Conservation efforts for the aye-aye are crucial for ensuring its survival. Various initiatives are in place to protect aye-aye populations, including habitat conservation, reforestation projects, and community education programs aimed at dispelling myths and promoting coexistence. Captive breeding programs in zoos and research institutions also play a role in preserving genetic diversity and raising awareness about this unique primate.

In conclusion, the aye-aye is one of the most mystifying and extraordinary animals in the world. Its distinctive appearance, specialized foraging behavior, and ecological significance make it a fascinating subject of study and conservation. Protecting the aye-aye and its habitat is essential for preserving Madagascar's unique biodiversity and ensuring the survival of this remarkable species.

8.   The Blobfish (Psychrolutes marcidus)

The blobfish, often dubbed the world's ugliest animal, is a deep-sea fish known for its gelatinous appearance and unique adaptations to life in the ocean's depths. Native to the deep waters off the coasts of Australia and New Zealand, the blobfish has become a symbol of the mysterious and often bizarre creatures that inhabit the ocean's dark and pressurized environments.

The blobfish's distinctive appearance is a result of its adaptation to the high-pressure, low-energy environment of the deep sea. At depths ranging from 600 to 1,200 meters (2,000 to 3,900 feet), where the pressure is dozens of times greater than at sea level, the blobfish's body is perfectly suited to its surroundings. Its flesh is primarily composed of a gelatinous mass with a density slightly less than water, allowing it to float above the seafloor with minimal energy expenditure. This adaptation is crucial for survival in an environment where food is scarce, and energy conservation is essential.

Out of water, the blobfish's gelatinous body collapses under its own weight, giving it the infamous "blob-like" appearance that has captured public imagination. However, in its natural habitat, the blobfish maintains a more typical fish shape, with its body supported by the surrounding water pressure. This drastic change in appearance when brought to the surface highlights the profound differences between deep-sea and surface environments.

The blobfish's diet primarily consists of deep-sea invertebrates and organic matter that drifts down from the upper layers of the ocean. It is a sedentary ambush predator, lying in wait for prey to come within reach. This passive feeding strategy is well-suited to the deep sea, where the energy cost of active hunting would be prohibitively high. The blobfish's large mouth allows it to consume a wide range of prey items, contributing to its success in this challenging environment.

Despite its unappealing appearance, the blobfish plays a vital role in the deep-sea ecosystem. As a scavenger and predator, it helps maintain the balance of species and contributes to the recycling of organic matter. The blobfish's ecological significance underscores the importance of preserving deep-sea habitats and understanding the creatures that inhabit them.

The blobfish gained widespread attention and notoriety in 2013 when it was voted the "World's Ugliest Animal" by the Ugly Animal Preservation Society, a campaign aimed at raising awareness about endangered and underappreciated species. This humorous yet impactful campaign brought the blobfish into the spotlight, highlighting the need for conservation efforts to protect deep-sea environments and the unique creatures that live there.

Despite its newfound fame, the blobfish faces several threats, primarily from deep-sea trawling and other fishing practices. These activities can inadvertently capture blobfish as bycatch, leading to population declines. The deep-sea habitats that blobfish rely on are also vulnerable to damage from bottom trawling and mining, which can disrupt the delicate ecosystems and reduce the availability of food sources.

Conservation efforts to protect the blobfish and other deep-sea species are crucial for maintaining the health and diversity of ocean ecosystems. These efforts include the establishment of marine protected areas, regulations on deep-sea fishing practices, and research initiatives to better understand deep-sea environments. Public awareness campaigns, like the one that brought attention to the blobfish, also play a vital role in promoting conservation and inspiring action to protect these remarkable creatures.

In conclusion, the blobfish is one of the most mystifying and misunderstood animals in the ocean. Its unique adaptations to the deep-sea environment, combined with its gelatinous appearance and ecological significance, make it a fascinating subject of study and conservation. Protecting the blobfish and its habitat is essential for preserving the rich biodiversity of the deep sea and ensuring that this extraordinary species continues to thrive.

9.   The Narwhal (Monodon monoceros)

The narwhal, often referred to as the "unicorn of the sea," is one of the most enigmatic and captivating marine mammals. Native to the Arctic waters of Canada, Greenland, Norway, and Russia, the narwhal is best known for its long, spiral tusk that protrudes from its head. This unique and mysterious feature has fascinated humans for centuries, inspiring myths and legends.

The narwhal's tusk is actually an elongated tooth that can grow up to 3 meters (10 feet) in length. It is most commonly found in males, though some females also possess smaller tusks. The tusk is spiraled counterclockwise and has a distinctive, smooth texture. It is made of ivory and is hollow for most of its length. The function of the tusk has been the subject of much speculation and scientific study. While it was once believed to be used for combat or breaking through ice, recent research suggests that the tusk is a sensory organ. It contains millions of nerve endings that can detect changes in water temperature, pressure, and salinity, helping narwhals navigate and find food in their icy habitat.

Narwhals are medium-sized whales, with males reaching lengths of up to 5.5 meters (18 feet) and females up to 4.5 meters (15 feet). They have a mottled gray and white coloration that provides excellent camouflage in their Arctic environment. Narwhals are highly social animals, often found in groups called pods that can range from a few individuals to several hundred. These pods engage in a variety of behaviors, including vocalizations, synchronized swimming, and diving.

The diet of narwhals primarily consists of fish, squid, and shrimp. They are known to dive to great depths, sometimes exceeding 1,500 meters (4,900 feet), to forage for food. Narwhals use echolocation to navigate and locate prey in the dark waters of the deep sea. Their ability to dive to such extreme depths and endure the cold temperatures of the Arctic Ocean is a testament to their remarkable adaptations.

Narwhals play a crucial role in the Arctic ecosystem and hold significant cultural importance for indigenous communities in the Arctic region. The Inuit people, for example, have a long history of hunting narwhals for their meat, blubber, and tusks, which are used for a variety of purposes, including tools, art, and traditional ceremonies. The sustainable hunting practices of indigenous communities have been vital for maintaining narwhal populations and preserving their cultural heritage.

Despite their adaptations to the harsh Arctic environment, narwhals face several threats that jeopardize their survival. Climate change is the most pressing threat, as it leads to the loss of sea ice, which narwhals rely on for breeding, feeding, and protection from predators. The reduction of sea ice also affects the availability of their prey and increases the risk of ship strikes and entanglement in fishing gear. Additionally, noise pollution from shipping and industrial activities can disrupt narwhal communication and navigation.

Conservation efforts are essential to protect narwhal populations and their Arctic habitat. International cooperation and agreements, such as those under the International Whaling Commission (IWC), play a critical role in regulating hunting practices and promoting sustainable management of narwhal populations. Research and monitoring programs are also crucial for understanding narwhal behavior, migration patterns, and population dynamics, which can inform conservation strategies.

In recent years, there has been a growing recognition of the need to address the impacts of climate change on Arctic species like the narwhal. Efforts to mitigate climate change through the reduction of greenhouse gas emissions, along with measures to protect critical habitats, are vital for ensuring the long-term survival of narwhals and other Arctic wildlife.

In conclusion, the narwhal is one of the most mystifying and remarkable animals in the ocean. Its unique tusk, incredible diving abilities, and cultural significance make it a species of immense scientific and cultural importance. Protecting the narwhal and its Arctic habitat is essential for preserving this extraordinary marine mammal and the delicate balance of the Arctic ecosystem.

Should we modify this page or shall we move on to the next

10. The Kakapo (Strigops habroptilus)

The kakapo, also known as the night parrot or owl parrot, is one of the most unique and endangered birds in the world. Native to New Zealand, the kakapo is a large, flightless parrot with distinctive nocturnal habits and a remarkable set of adaptations. Despite its charming appearance and fascinating behaviors, the kakapo faces critical threats to its survival, making it a key focus of intensive conservation efforts.

The kakapo is notable for its striking appearance. It is the heaviest parrot species in the world, with males weighing up to 4 kilograms (8.8 pounds). Its plumage is a blend of mossy green and yellow, which provides excellent camouflage against the forest floor and foliage. This cryptic coloration helps the kakapo remain hidden from predators, a crucial adaptation given its inability to fly.

One of the most unusual aspects of the kakapo is its nocturnal and ground-dwelling lifestyle. Unlike most parrots, which are diurnal and arboreal, the kakapo is primarily active at night and spends much of its time on the ground. It has strong legs and large, clawed feet, which it uses to navigate through its forest habitat with agility. The kakapo's nocturnal habits and keen sense of smell help it locate food, which consists of a variety of native plants, seeds, fruits, and even bark.

The kakapo's reproductive behavior is equally unique. It is one of the few bird species that engage in a lek mating system. During the breeding season, males gather in communal areas known as leks, where they compete for the attention of females through a series of elaborate displays and calls. The males produce a distinctive low-frequency booming call, which can be heard over long distances and is used to attract females to their display sites. After mating, the female kakapo lays her eggs in a secluded nest on the ground and raises the chicks alone.

Despite its fascinating adaptations, the kakapo is critically endangered, with a population of fewer than 250 individuals. The primary threats to the kakapo's survival include habitat loss, introduced predators, and a low reproductive rate. The arrival of humans in New Zealand and the subsequent introduction of mammalian predators such as rats, stoats, and cats have had a devastating impact on kakapo populations. These predators, to which the kakapo has no natural defenses, prey on eggs, chicks, and even adult birds.

In response to the severe decline in kakapo numbers, intensive conservation efforts have been implemented. The Kakapo Recovery Program, launched in the 1990s, involves a combination of predator control, habitat restoration, and a managed breeding program. The remaining kakapo population has been relocated to predator-free offshore islands, where they are closely monitored and protected by conservationists.

The success of the Kakapo Recovery Program has been remarkable, with the population steadily increasing thanks to active management and innovative conservation techniques. Artificial insemination, hand-rearing of chicks, and the use of technology to monitor and support breeding efforts have all contributed to the recovery of the kakapo. Public awareness campaigns and community involvement have also played a crucial role in garnering support for kakapo conservation.

Despite these successes, the kakapo remains one of the most endangered birds in the world, and ongoing efforts are essential to ensure its long-term survival. Research into the kakapo's genetics, health, and behavior continues to inform conservation strategies, while habitat protection and predator control remain top priorities. In conclusion, the kakapo is one of the most mystifying and extraordinary animals in the world. Its unique adaptations, nocturnal habits, and critically endangered status make it a species of immense scientific and conservation importance. Protecting the kakapo and ensuring its survival is not only crucial for preserving biodiversity but also serves as a testament to the power of dedicated conservation efforts and the importance of protecting our natural heritage.

Chapter 4: Odd Phenomena

1. Bioluminescent Bays

Bioluminescent bays, also known as bio bays, are among the most enchanting and mysterious natural wonders on Earth. These luminous bodies of water, where the sea glows with a magical blue-green light, captivate visitors and scientists alike. Found in a few select locations around the world, bioluminescent bays are a testament to the remarkable phenomena that occur in nature.

Bioluminescence is the production and emission of light by living organisms. In the case of bioluminescent bays, the glow is primarily caused by tiny marine plankton called dinoflagellates. These microscopic organisms emit light when they are disturbed, creating a dazzling display that illuminates the water with every movement. The light produced by dinoflagellates is a result of a chemical reaction involving luciferin (a light-emitting molecule) and luciferase (an enzyme). This reaction is triggered by mechanical stress, such as waves, fish swimming, or even a paddle from a kayak.

One of the most famous bioluminescent bays in the world is Mosquito Bay, located on the island of Vieques in Puerto Rico. Often considered the brightest bioluminescent bay, Mosquito Bay is renowned for its intense glow, which can light up the entire bay on a dark night. The high concentration of dinoflagellates in this bay makes it an ideal spot for experiencing the magic of bioluminescence.

Another notable location is Toyama Bay in Japan, known for its seasonal displays of bioluminescent firefly squid. Every spring, millions of these tiny squid gather near the surface of the water to spawn, creating a spectacular light show. The bioluminescent glow of the firefly squid is a mesmerizing sight, attracting tourists and researchers from around the world.

Bioluminescent bays are not only beautiful but also ecologically significant. The light produced by dinoflagellates serves as a defense mechanism against predators. The sudden flash of light can startle and confuse predators, allowing the plankton to escape. Additionally, bioluminescence plays a role in communication and mating for some marine species. The glow of bioluminescent organisms can attract mates or signal danger to other members of the same species.

The phenomenon of bioluminescence has a profound impact on local tourism. Bioluminescent bays attract thousands of visitors each year, eager to witness the natural light show. Kayaking and boat tours are popular activities, allowing tourists to immerse themselves in the glowing waters. The economic benefits of bioluminescent tourism are significant, providing income and employment opportunities for local communities.

However, bioluminescent bays face several conservation challenges. Human activities, such as coastal development, pollution, and overfishing, can negatively impact the delicate

ecosystems that support bioluminescence. Light pollution from nearby towns and
resorts can also diminish the visibility of the bioluminescent glow, reducing the magical experience for visitors. Additionally, climate change poses a threat to these unique ecosystems. Changes in sea temperature, salinity, and pH levels can affect the population and behavior of dinoflagellates, potentially reducing the intensity and frequency of bioluminescent displays.

Conservation efforts are crucial to protect bioluminescent bays and ensure their continued existence. This includes implementing measures to reduce pollution, regulate coastal development, and control light pollution. Educating tourists and local communities about the importance of preserving these natural wonders is also essential. Sustainable tourism practices, such as limiting the number of visitors and promoting eco-friendly activities, can help minimize the impact on these fragile ecosystems.

Scientific research plays a vital role in understanding and protecting bioluminescent bays. Ongoing studies aim to uncover the intricacies of bioluminescence and the environmental factors that influence it. By gaining a deeper understanding of the biological and ecological aspects of bioluminescence, scientists can develop more effective conservation strategies and mitigate the impact of human activities on these unique habitats.

In conclusion, bioluminescent bays are among the most captivating and mysterious natural phenomena in the world. The enchanting glow of the water, caused by the bioluminescence of dinoflagellates, creates a magical experience that attracts visitors from all corners of the globe. However, the delicate ecosystems that support bioluminescence are under threat from human activities and environmental changes. Protecting and preserving bioluminescent bays is essential for maintaining their ecological significance and ensuring that future generations can continue to marvel at their beauty. Through conservation efforts, scientific research, and sustainable tourism practices, we can help safeguard these natural wonders and keep their magical glow alive.

2. The Taos Hum

The Taos Hum is one of the most perplexing and widely reported auditory phenomena in the world. Named after the town of Taos in New Mexico, where it was first widely reported in the early 1990s, the Taos Hum is described as a persistent low-frequency noise that only a small percentage of the population can hear. This mysterious hum has sparked numerous theories and investigations, yet its origin remains elusive.

The Taos Hum is typically described as a faint, droning sound, akin to the hum of an engine idling in the distance. It is generally heard indoors, and reports indicate that it is most noticeable at night or in quiet environments. People who hear the hum, known as "hearers," often describe it as an annoying and sometimes distressing noise that can interfere with sleep and concentration. The hum is usually reported to be around a frequency of 56 Hz, but it can vary in pitch and intensity.

The phenomenon is not limited to Taos; similar hums have been reported in various locations worldwide, including Bristol in the United Kingdom, Largs in Scotland, and Windsor in Canada. Each of these locations has its own version of the hum, with slight variations in sound characteristics and the percentage of the population affected.

Several theories have been proposed to explain the Taos Hum, ranging from environmental and industrial sources to more unconventional ideas. One of the most common theories is that the hum is caused by low-frequency sound waves generated by industrial equipment, such as gas pipelines, electrical transformers, or large ventilation systems. These sound waves can travel long distances and penetrate buildings, making them audible to some individuals.

Another theory suggests that the hum could be related to natural geological processes, such as the movement of tectonic plates or volcanic activity. These natural phenomena can produce low-frequency vibrations that might be perceived as a hum. Additionally, some researchers have proposed that the hum could be a form of tinnitus, a condition where individuals hear sounds that are not present in the external environment. However, this theory does not fully explain why only certain locations and a small percentage of people are affected.

Scientific investigations into the Taos Hum have yielded mixed results. In 1993, a team of researchers from the University of New Mexico conducted a study to identify the source of the hum in Taos. They used various instruments to measure sound levels and frequencies but were unable to find a definitive source. Some researchers suggest that the hum may be a combination of multiple factors, both environmental and physiological.

The psychological and physiological effects of the Taos Hum on hearers are significant. Many report feelings of frustration, anxiety, and sleep disturbances due to the persistent noise. Some individuals have even relocated to escape the hum, only to find that it follows them to new locations. This has led to increased interest in finding effective ways to mitigate the impact of the hum on those affected.

Cultural and local responses to the Taos Hum vary. In some communities, the hum has become a part of local folklore, with various legends and stories explaining its origin. In Taos, the hum has attracted the attention of researchers, journalists, and tourists, adding to the town's mystique. Local authorities have also taken steps to address concerns from residents, including funding studies and holding public meetings to discuss the phenomenon.

81

In conclusion, the Taos Hum is one of the most mystifying and widely reported auditory phenomena in the world. Despite numerous theories and investigations, its origin remains unknown. The hum continues to intrigue and perplex scientists, hearers, and the general public. Understanding the Taos Hum and finding ways to alleviate its impact on those affected is an ongoing challenge that underscores the complexity and mystery of our natural world.

3.   The Sailing Stones of Death Valley

The sailing stones, also known as sliding rocks or moving rocks, of Death Valley are one of the most enigmatic and fascinating natural phenomena in the world. Located in Racetrack Playa, a remote dry lakebed in Death Valley National Park, California, these stones mysteriously move across the desert floor, leaving long trails behind them. This phenomenon has puzzled scientists and visitors for decades, prompting numerous studies and theories to explain their movement.

Racetrack Playa is a nearly flat, dry lakebed surrounded by mountains. The playa's surface is composed of fine clay, which becomes extremely slick and muddy when wet. The stones on Racetrack Playa vary in size, with some weighing several hundred pounds. Despite their weight, these stones have been observed to travel significant distances, leaving straight or curved tracks in the clay surface.

For many years, the movement of the sailing stones remained a mystery. Early theories suggested that strong winds pushed the stones across the playa. However, the sheer size and weight of some stones made this explanation seem unlikely. Other hypotheses included magnetic fields, earthquakes, and even supernatural forces, but none of these theories provided a satisfactory explanation.

The mystery of the sailing stones was finally solved in 2014, thanks to the efforts of a team of researchers from the Scripps Institution of Oceanography and NASA. Using time-lapse photography, GPS tracking, and weather data, the researchers discovered that a unique combination of environmental conditions was responsible for the movement of the stones.

The key to understanding the sailing stones lies in the interplay between water, ice, and wind. During the winter, rainwater accumulates on the surface of Racetrack Playa, forming a shallow, ephemeral pond. When temperatures drop at night, this water freezes, creating a thin layer of ice. As the sun rises and temperatures increase, the ice begins to melt and break into large floating panels. A gentle breeze then pushes these ice panels across the playa, carrying the stones with them. The ice acts as a raft, reducing friction and allowing the stones to slide across the wet, slippery clay surface. This process can create the long, distinct trails that have puzzled observers for so long.

The discovery of this mechanism not only solved the mystery of the sailing stones but also highlighted the complexity and beauty of natural processes. It demonstrated how seemingly simple elements—water, ice, and wind—can interact in unexpected ways to produce extraordinary results.

Racetrack Playa and its sailing stones have become a popular destination for tourists and geology enthusiasts. Visitors to the playa are often struck by the stark beauty of the landscape and the eerie tracks left by the moving stones. However, it is important to protect this delicate environment from damage. The National Park Service has implemented guidelines to minimize human impact, including prohibiting the removal of stones and limiting vehicle access to designated areas.

The scientific significance of the sailing stones extends beyond Death Valley. Understanding the mechanisms behind their movement can provide insights into similar phenomena on other planets. For example, researchers have suggested that similar processes involving ice and wind might occur on Mars, where evidence of past water and ice has been found. Studying the sailing stones of Racetrack Playa can help scientists develop models to explain surface features on Mars and other planetary bodies.

In conclusion, the sailing stones of Death Valley are one of the most mystifying and captivating natural phenomena in the world. The discovery of the environmental conditions that drive their movement has solved a long-standing mystery and highlighted the intricate interplay of natural forces. Protecting Racetrack Playa and continuing to study this remarkable phenomenon are essential for understanding and appreciating the complexity and beauty of our planet.

4. Ball Lightning

Ball lightning is one of the most mysterious and elusive atmospheric phenomena observed by humans. Described as a glowing, spherical object that appears during thunderstorms, ball lightning has fascinated and perplexed scientists and eyewitnesses for centuries. Despite numerous reports and studies, the exact nature and cause of ball lightning remain uncertain, making it one of the most intriguing natural phenomena.

Ball lightning typically appears as a glowing orb, ranging in size from a few centimeters to several meters in diameter. The color of the ball can vary, with reports of red, orange, yellow, blue, and white orbs. It usually lasts for a few seconds to a few minutes, moving erratically through the air before either fading away or disappearing with a loud bang. Some reports describe the ball passing through solid objects, such as windows or walls, without causing damage, while others mention the orb leaving burn marks or causing fires.

Historical accounts of ball lightning date back to ancient times, with one of the earliest recorded instances described by the Greek philosopher Aristotle in the 4th century BCE. Since then, numerous eyewitness reports have emerged from all over the world, contributing to the rich lore and mystery surrounding the phenomenon.

Despite the abundance of eyewitness accounts, scientific understanding of ball lightning is limited due to its unpredictable and fleeting nature. Capturing ball lightning in controlled laboratory conditions has proven to be extremely challenging, which has hindered comprehensive scientific study. However, several theories have been proposed to explain the phenomenon.

One of the leading theories suggests that ball lightning is a form of plasma, a state of matter composed of electrically charged particles. According to this theory, ball lightning is created when a lightning strike vaporizes certain elements in the air or ground, forming a plasma ball that remains stable for a short period. This theory is supported by the observation that ball lightning often occurs during thunderstorms and is sometimes seen in conjunction with regular lightning strikes.

Another theory posits that ball lightning is caused by the discharge of energy from a high-voltage electrical field. In this scenario, the electrical field ionizes the air, creating a luminous ball of ionized gas. This theory is consistent with reports of ball lightning appearing near power lines, electrical equipment, or other sources of strong electrical fields.

A more recent hypothesis suggests that ball lightning is the result of a chemical reaction involving silicon. According to this theory, lightning strikes the ground and vaporizes silica, creating a cloud of silicon nanoparticles. These particles then react with oxygen in the air, releasing energy in the form of light and heat, which forms the glowing ball of ball lightning.

Despite these theories, no single explanation has been universally accepted, and the exact mechanism behind ball lightning remains a topic of active research and debate. Advances in technology, such as high-speed cameras and sophisticated sensors, have provided new opportunities for studying ball lightning and may eventually lead to a definitive understanding of this enigmatic phenomenon. The challenges in studying ball lightning are compounded by its rarity and unpredictability. Eyewitness reports, while valuable, are often inconsistent and difficult to verify scientifically. As a result, researchers rely heavily on anecdotal evidence and theoretical models to piece together a coherent explanation.

The technological implications of understanding ball lightning are significant. Insights into the formation and behavior of ball lightning could lead to advancements in energy storage, plasma technology, and electrical engineering. Additionally, studying ball lightning can improve our understanding of atmospheric physics and the complex interactions between electrical fields and matter.

In conclusion, ball lightning is one of the most mystifying and elusive phenomena in the natural world. Despite centuries of reports and numerous theories, its exact nature remains a mystery. Continued research and technological advancements hold the promise of unlocking the secrets of ball lightning, shedding light on this captivating and enigmatic phenomenon.

5.   The Hessdalen Lights

The Hessdalen Lights, an unexplained light phenomenon observed in the Hessdalen Valley of Norway, have intrigued scientists and enthusiasts for decades. These mysterious lights, which have been recorded since the early 1980s, display a range of behaviors and characteristics that defy easy explanation. The Hessdalen Lights are one of the most persistent and well-documented examples of unexplained atmospheric phenomena in the world.

The Hessdalen Valley, located in central Norway, is a remote and sparsely populated area. The lights are typically seen in the night sky and have been reported in various colors, including white, yellow, and red. They appear as floating orbs or streaks of light, often moving slowly across the valley before disappearing. The lights can last from a few seconds to over an hour, and their brightness can vary significantly.

The Hessdalen Lights exhibit several distinct behaviors that add to their mystery. They can hover in place, move rapidly, or oscillate back and forth. Sometimes they appear to follow specific patterns or form geometric shapes. Witnesses have also reported seeing multiple lights simultaneously, interacting with each other in complex ways. These unpredictable movements and formations have made the Hessdalen Lights a fascinating subject for study.

Numerous theories have been proposed to explain the Hessdalen Lights, ranging from natural atmospheric phenomena to more unconventional ideas. One of the leading scientific hypotheses suggests that the lights are a form of plasma, created by the ionization of gases in the atmosphere. According to this theory, the unique geological and meteorological conditions in the Hessdalen Valley may produce a natural electrical discharge, similar to ball lightning. This discharge could ionize the surrounding air, creating the glowing orbs and streaks observed in the sky.

Another theory posits that the lights are caused by piezoelectricity, a type of electrical charge generated by certain materials under mechanical stress. The Hessdalen Valley is rich in quartz, a mineral known for its piezoelectric properties. Geological activity, such as tectonic movements, could create stress in the quartz deposits, generating an electrical charge that ionizes the air and produces the lights.

In addition to these scientific explanations, some researchers have explored the possibility that the Hessdalen Lights are related to unidentified flying objects (UFOs) or extraterrestrial activity. While this idea remains highly speculative and controversial, it continues to capture the imagination of many enthusiasts and has contributed to the enduring allure of the Hessdalen Lights.

The Hessdalen Lights have attracted significant scientific attention over the years. In the early 1980s, a team of researchers from the Østfold University College in Norway established Project Hessdalen to study the phenomenon. The project involved extensive fieldwork, including the installation of monitoring equipment and the collection of eyewitness reports. Despite numerous observations and data collection efforts, the exact cause of the Hessdalen Lights remains elusive.

One of the key findings from Project Hessdalen was the identification of several environmental factors that may contribute to the phenomenon. These factors include the presence of metallic ores, such as iron and copper, in the valley's geology, as well as specific weather conditions that may enhance the likelihood of electrical discharges. However, these findings have not led to a definitive explanation, and the lights continue to be a subject of active research.

The impact of the Hessdalen Lights on local tourism and culture has been significant. The valley has become a popular destination for tourists and researchers interested in witnessing the lights firsthand. Local businesses have capitalized on the phenomenon, offering guided tours and accommodations for visitors. The lights have also inspired various cultural works, including documentaries, books, and art installations.

In conclusion, the Hessdalen Lights are one of the most mystifying and well-documented unexplained phenomena in the world. Their persistent and unpredictable nature has defied easy explanation, prompting numerous theories and extensive scientific research. While the exact cause of the lights remains unknown, their study continues to provide valuable insights into the complex interactions between geology, meteorology, and atmospheric physics. Protecting the Hessdalen Valley and continuing to study this remarkable phenomenon are essential for unlocking the secrets of the Hessdalen Lights and advancing our understanding of the natural world.

6. The Rolling Stones of Death Valley

One of the world's most mysterious and fascinating natural phenomena is the Death Valley Rolling Stones, also known as 'Sailing Stones' in English. These stones seem to move by themselves across the dry bed of a lake called Racetrack Playa, leaving behind long trails in the dry, cracked soil. This phenomenon has aroused the curiosity of scientists, tourists and fans of natural mysteries for decades.

The characteristics and dynamics of this phenomenon are as fascinating as the phenomenon itself:

Environment: Racetrack Playa is an ancient dry lake located in California's Death Valley. This plain, about 4 kilometres long and 2 kilometres wide, is covered by a layer of dry, cracked mud that forms when the few rains that fall in the region quickly evaporate under the scorching desert sun.

The Stones: The stones that move across the surface of the playa vary in size, from small pebbles to large boulders weighing hundreds of kilos. Despite their weight, these stones manage to move, leaving visible and often sinuous tracks on the ground.

Wakes: The trails left by the stones can be up to hundreds of metres long and vary in direction and length. Some trails are straight, others curvilinear, and often the tracks intersect each other, creating an intricate pattern on the ground.

Hypotheses and Studies: For many years, the exact cause of the movement of the stones remained a mystery. Several hypotheses have been proposed, including strong winds, slippery ground caused by rain and even magnetic forces. However, none of these explanations could fully explain the phenomenon.

Discovery of the Cause: In 2014, a team of researchers from the Scripps Institute of Oceanography finally unravelled the mystery through a detailed experiment. Using time-lapse cameras and GPS instruments attached to the stones, the researchers discovered that the movement of the stones is caused by a combination of rain, night frost and light winds. During the winter, rain forms a thin layer of water on the surface of the playa. At night, this water freezes, forming thin but strong ice sheets. When the morning sun partially melts the ice, the sheets break into smaller pieces that, pushed by light winds, can push the stones slowly across the muddy surface.

Implications: The discovery highlighted the importance of climatic and environmental conditions in creating seemingly inexplicable natural phenomena. It also demonstrated how modern technology can be used to solve mysteries that have fascinated mankind for decades.

The Death Valley Rolling Stones are an extraordinary example of how nature can create complex and fascinating phenomena that defy our understanding. Their discovery and scientific explanation have not only solved a long-standing mystery but have also strengthened our appreciation for the beauty and complexity of our planet.

7.   Lake Hillier

Lake Hillier, located on Middle Island off the coast of Western Australia, is one of the world's most enigmatic and fascinating natural phenomena due to its bright pink colour. This salt lake, about 600 metres long, is surrounded by dense eucalyptus forest and the blue waters of the Pacific Ocean, creating an extraordinary colour contrast that attracts the attention of scientists and tourists alike.

The characteristics and origin of Lake Hillier's distinctive colour are the subject of scientific interest and research:

Colour: The pink colour of Lake Hillier is permanent and does not change even if the water is taken and transferred to another container. This distinguishes it from other salt lakes that change colour depending on weather conditions or the time of year.

Causes of Colour: The exact cause of Lake Hillier's pink colour is not yet fully understood, but the most accepted explanation involves the presence of microorganisms. Two species of microorganisms in particular, the algae Dunaliella salina and the bacterium Halobacterium, are known to thrive in extreme saline environments. Dunaliella salina produces a red carotenoid pigment, beta-carotene, which is believed to contribute to the pink colour. Halobacterium, which possess a red pigment known as bacterioruberin, may also contribute to the phenomenon.

Salinity: Lake Hillier is highly saline, with a salt concentration similar to that of the Dead Sea. This extreme environment favours the proliferation of halophilic organisms, such as those mentioned above, which are able to survive and thrive in conditions that would be lethal to most other life forms.

Discovery and Exploration: Lake Hillier was discovered in 1802 by the British navigator Matthew Flinders, who noted its extraordinary colouring in his logbook. Since then, the lake has become a destination of scientific and tourist interest. However, due to its remote location on a protected island, access to the lake is limited and is mainly via helicopter excursions.

Conservation: Middle Island is part of the Recherche Archipelago and is protected by Australian environmental laws. The conservation of Lake Hillier's unique ecosystem is critical to preserving its natural beauty and continuing scientific research into its unique characteristics.

Global comparisons: Although Lake Hillier is one of the best known pink lakes in the world, it is not the only one. Other pink lakes are found in different parts of the globe, such as Lake Retba in Senegal and Lake Dusty Rose in Canada. These lakes share similar characteristics, such as high salinity and the presence of pigment-producing microorganisms, but Lake Hillier remains unique because of its consistent colouring and the combination of factors that contribute to its distinctive appearance.

Lake Hillier represents one of the most fascinating examples of how nature can create spectacles of beauty and mystery. Its extraordinary colour and distinctive ecology offer unique opportunities for scientific research and for appreciating the diversity and complexity of our planet's natural environments.

8.   The Crooked Forest

The Crooked Forest, known locally as 'Krzywy Las,' is a strange and fascinating forest area located near the town of Gryfino in north-western Poland. This forest consists of around 400 pine trees that have a unique and mysterious feature: the trunks of the trees grow with a 90-degree bend at the base, then straighten upwards. This phenomenon has raised numerous questions and hypotheses about its origin, making the Foresta Storta a world-famous botanical enigma.

The characteristics and theories surrounding the Foresta Storta are as fascinating as the phenomenon itself:

Appearance: The pines of the Foresta Storta grow with a sharp, uniform curve at the base, bending northwards before straightening out. The curve of the trunks starts a few centimetres above the ground and reaches a length of about one metre before growing vertically again.

Age and Origin: The trees are estimated to have been planted in the 1930s, but the precise origin of the phenomenon remains unknown. The abnormal growth is unique to this area and is not found in other surrounding forests.

Theories of Origin: Numerous theories have been proposed to explain the curvature of the trunks. These include:

Human intervention: One of the most widely accepted theories suggests that the bending was caused by deliberate human intervention. It is believed that the trees were bent intentionally during their growth phase for commercial purposes, such as the production of timber for curved furniture or boat hulls.

Damage During Growth: Another theory proposes that the trees were damaged by mechanical means, such as war wagons during World War II, and that subsequent growth produced the characteristic bend.

Natural Phenomena: Some suggest that unique environmental conditions, such as heavy snowfall or windstorms, caused the trunks to deform.

Scientific Research: Despite numerous theories, there is still no definitive explanation for the abnormal growth of the Stunted Forest. The lack of detailed historical documentation makes it difficult to verify the proposed hypotheses. The forest continues to be an object of study for botanists and researchers trying to solve the mystery.

Tourist Attraction: The Foresta Storta has become a popular tourist destination due to its uniqueness and the mystery surrounding it. Every year, numerous visitors come to Gryfino to see the bent trees in person and to take photographs of this unique natural phenomenon.

Conservation: Protecting the Crooked Forest is crucial to preserving this unique phenomenon. Local authorities and conservation groups work to ensure that the area remains intact and accessible to visitors, avoiding damage to the trees.

The Foresta Storta is an intriguing example of how nature can surprise and fascinate with unexpected and unexplainable phenomena. The combination of natural beauty and mystery makes this forest a place of interest for both nature lovers and the curious in search of enigmas to solve. Although many theories have been proposed, the true cause of the curvature of the trees in the Foresta Storta remains one of the great botanical mysteries of our time.

9. The Hessdalen Lights

The Hessdalen Lights are a mysterious atmospheric phenomenon occurring in the Hessdalen valley in Norway. These unexplained lights have been observed regularly since the 1980s, attracting the attention of scientists, researchers and onlookers around the world. The lights appear as luminous spheres of various sizes and colours floating in the sky, often moving in unconventional ways and sometimes remaining visible for hours.

The characteristics and theories surrounding the Hessdalen Lights are just as fascinating as the phenomenon itself:

Appearance: The Hessdalen Lights can appear as luminous spheres, streaks of light or even intermittent flashes. The colours vary from white to yellow, red, blue and green. Their intensity and duration can vary greatly, with some lights disappearing quickly and others remaining visible for several minutes or even hours.

Frequency and Location: The lights have been sighted mainly in the Hessdalen valley, a rural and sparsely populated area in central Norway. The frequency of sightings peaked in the 1980s, with over

20 observations per month, but the phenomenon continues to occur, albeit less frequently, today.

Scientific Theories: Several theories have been proposed to explain the Hessdalen Lights, but none have been definitively proven. Scientific explanations include:

Plasma: Some scientists speculate that the lights are caused by plasma, an ionised gas that can emit light. This could form due to electrical interactions in the subsurface, such as the oxidation of metal ores.

Electromagnetic Radiation: Another theory suggests that the lights may be the result of electromagnetic radiation from natural sources, such as electric fields produced by lightning or geomagnetic storms.

Natural Gases: Some researchers believe the phenomenon may be due to emissions of natural gases, such as methane, which spontaneously ignite when they come into contact with air.

Research Projects: The Hessdalen Valley has been the subject of numerous scientific studies to try to understand the phenomenon. Since 1983, the Hessdalen Project has monitored and documented the lights using cameras, radar, and other scientific instruments. These studies have collected a great deal of data, but the phenomenon still remains shrouded in mystery.

International Interest: The Lights of Hessdalen have attracted the attention of international researchers and fans of unexplained phenomena. The valley has become a place of pilgrimage for those hoping to witness this mysterious light show in person. In addition, the phenomenon has inspired documentaries, books and scientific articles.

Implications and Importance: Studying the Hessdalen Lights could not only solve a local mystery but could also provide new information on atmospheric and geophysical processes. Understanding this phenomenon could have implications for

95

research on lightning, natural gas emissions and other atmospheric electrical phenomena.

The Hessdalen Lights represent one of the most enigmatic natural phenomena of our time. Their beauty and mystery continue to inspire and challenge the scientific community and the public. While research continues, the Hessdalen Lights remain a fascinating example of how nature can still hold surprises and phenomena that challenge our understanding.

10. The Moeraki Stones

The Moeraki Stones are another extraordinary natural phenomenon that fascinates visitors and scientists alike. These rock formations, located along the coast of Koekohe Beach on New Zealand's South Island, are large spherical boulders that appear to have been carved by a giant hand and scattered along the beach. The stones have sparked numerous theories about their origin and composition, making them a tourist attraction and an interesting subject for geological study.

The characteristics and theories surrounding the Moeraki Stones are as fascinating as the phenomenon itself:

Appearance: The Moeraki Stones are known for their almost perfectly spherical shape and impressive size. Some of them can reach a diameter of more than two metres and weigh several tonnes. The surfaces of the stones often have cracks and lines that make them even more fascinating.

Composition: The stones consist of a core of mud, clay and silt, surrounded by a hard layer of limestone. This outer layer was formed through a natural cementing process that compacted the marine sediments over millions of years.

Formation: The most widely accepted theory of the formation of the Moeraki Stones is that of concretion. Concretions form when minerals dissolved in water fill the spaces between sediment grains, hardening over time. This process took place below sea level millions of years ago when sediment particles compacted together, forming these rocky spheres.

Erosion: Over time, coastal erosion processes have exposed the Moeraki Stones, which are now visible on the beach. Waves and wind continue to shape these stones, making them increasingly visible and accessible to visitors.

Maori legends: The Moeraki Stones also have an important cultural significance for the Maori, the indigenous people of New Zealand. According to Maori legend, the stones represent the remains of baskets of food, pumpkins and sweet potatoes washed ashore by the great canoe Arai-te-uru, which capsized on the coast.

Scientific Research: The Moeraki Stones have been the subject of numerous geological studies to better understand the processes of concretionary formation and coastal erosion. Researchers continue to examine these stones to gather valuable data on climate change and geological evolution in the region.

Tourist Attraction: Today, the Moeraki Stones are a popular tourist attraction. Visitors from all over the world come to Koekohe Beach to admire these unique formations and take photographs. Their unusual shape and fascinating history help make this site one of the most visited places in New Zealand.

Conservation: The protection of the Moeraki Stones is crucial to preserving this natural heritage. Local authorities and conservation groups work to ensure that tourist access does not damage the stones and that the surrounding beaches remain clean and pristine.

The Moeraki Stones are an extraordinary example of how natural processes can create unique and fascinating geological formations. These spherical boulders not only offer an amazing visual spectacle, but also a window into Earth's geological past. The combination of natural beauty, cultural significance and scientific

importance makes the Moeraki Stones a phenomenon to be
preserved and studied for future generations.

Chapter 5: Curious Human Anomalies

1.   The Man with the Golden Arm

James Harrison, famously known as the "Man with the Golden
Arm," has a unique and life-saving gift. His extraordinary blood
plasma has saved millions of lives, particularly newborns suffering
from Rhesus disease, a severe form of anemia. Harrison's
remarkable contribution to medical science and the lives he has
touched has made him a true hero and a symbol of human kindness
and generosity.

James Harrison was born in 1936 in Australia. His journey into the
medical world began when he was just 14 years old. Harrison
underwent major chest surgery, requiring a significant amount of
blood. This experience instilled in him a deep appreciation for
blood donors and motivated him to become one himself once he
turned 18. Little did he know, his blood would turn out to be
exceptionally rare and valuable.

Shortly after he started donating blood, doctors discovered that
Harrison's blood contained a rare antibody. This antibody could be
used to create a life-saving treatment for Rhesus disease, also

known as hemolytic disease of the newborn. Rhesus disease occurs when a pregnant woman's immune system attacks the blood cells of her unborn baby. This condition can lead to severe anemia, brain damage, or even death in the fetus or newborn.

The antibody in Harrison's blood is known as Anti-D. It is used to produce an injection called Anti-D immunoglobulin, which prevents the immune system of Rh-negative mothers from attacking their Rh-positive babies' red blood cells. Before this treatment was developed, thousands of babies in Australia died each year from Rhesus disease. Harrison's plasma has been used to create more than 3 million doses of Anti-D, significantly reducing the incidence of the disease and saving countless lives.

Harrison's contributions are immense. He has donated blood over 1,100 times, almost every week for over 60 years, until he reached the age limit for donors in 2018. His donations are estimated to have saved the lives of over 2.4 million babies. Harrison's commitment and selflessness have made a profound impact on many families, earning him the nickname "Man with the Golden Arm."

The scientific explanation behind Harrison's unique blood lies in his immune system's ability to produce the Anti-D antibody. Researchers believe this rare antibody may have developed as a response to the numerous blood transfusions Harrison received during his surgery as a teenager. The presence of this antibody in his blood has provided a crucial tool in combating Rhesus disease, showcasing the importance of individual contributions to medical advancements.

James Harrison's dedication to blood donation has been widely recognized and celebrated. He has received numerous awards and honors, including the Medal of the Order of Australia in 1999. His story has also raised awareness about the importance of blood donation and inspired many others to become regular donors.

In addition to his medical contributions, Harrison's story is a testament to the power of human kindness and the impact one individual can have on the world. His willingness to give selflessly, week after week, year after year, highlights the extraordinary capacity for compassion and generosity that exists within us all.

James Harrison's legacy continues to inspire and remind us of the importance of blood donation. While his remarkable journey as a blood donor has come to an end, the lives he has saved and the families he has helped will always be a testament to his incredible generosity. The "Man with the Golden Arm" has left an indelible mark on medical history and on the hearts of millions around the world.

2. The Human Calculator

Shakuntala Devi, known as the "Human Calculator," astonished the world with her extraordinary mathematical abilities. Born in Bangalore, India, in 1929, Devi's prodigious talent for mental arithmetic emerged at an early age. Her incredible speed and accuracy in performing complex calculations without any mechanical aid earned her international fame and a place in the Guinness Book of World Records.

Shakuntala Devi's mathematical journey began when she was just three years old. Her father, a circus performer, noticed her uncanny ability to memorize numbers and perform mental calculations. Recognizing her potential, he nurtured her talent, and soon she was performing complex arithmetic feats at public shows. By the age of six, she had become a child prodigy, captivating audiences with her demonstrations of mental mathematics.

Devi's reputation as a mathematical genius continued to grow as she traveled the world, performing at universities and on television. Her most famous feat occurred in 1980, when she multiplied two 13-digit numbers in just 28 seconds, a record that earned her a place in the Guinness Book of World Records. This remarkable achievement highlighted her extraordinary cognitive abilities and brought her widespread acclaim.

Shakuntala Devi's talents extended beyond arithmetic. She was also a gifted astrologer and authored several books on mathematics, puzzles, and astrology. Her book "The World of Homosexuals," published in 1977, was one of the first studies on homosexuality in India, showcasing her versatility and intellectual curiosity. Despite her diverse interests, it was her mathematical prowess that captivated the world and cemented her legacy as the "Human Calculator."

The scientific community has long been fascinated by Devi's abilities, and researchers have sought to understand the cognitive mechanisms behind her exceptional skills. Some scientists believe that her abilities may be linked to her brain's unique neural architecture, which allowed her to process numerical information at an extraordinary speed. Others suggest that her talent was the result of intense practice and a deep understanding of mathematical principles from an early age.

Shakuntala Devi's impact on mathematics and education is profound. She inspired countless individuals to appreciate the beauty and power of mathematics, and her demonstrations of mental arithmetic showcased the potential of the human mind. Her work also highlighted the importance of nurturing young talents and providing opportunities for intellectual growth.

Devi's life and achievements continue to be celebrated today. In 2020, a biographical film titled "Shakuntala Devi" was released, starring Vidya Balan in the titular role. The film brought her story to a new generation, emphasizing her contributions to mathematics and her indomitable spirit.

In addition to her mathematical feats, Shakuntala Devi's legacy includes her contributions to education. She conducted numerous workshops and seminars, sharing her techniques and methods for improving mental arithmetic skills. Her approach to teaching mathematics was rooted in making the subject accessible and enjoyable, helping to demystify complex concepts and inspire a love for learning.

Shakuntala Devi's life is a testament to the extraordinary potential of the human mind. Her ability to perform complex calculations with astonishing speed and accuracy remains unparalleled, and her contributions to mathematics and education have left an indelible mark on the world. As the "Human Calculator," she demonstrated that with talent, dedication, and a passion for learning, one can achieve remarkable feats and inspire others to reach their full potential.

3.   The Tree Man

Dede Koswara, known as the "Tree Man," gained global attention due to his rare and debilitating skin condition called epidermodysplasia verruciformis (EV). This rare genetic disorder causes wart-like lesions and growths that resemble tree bark to spread across the body. Koswara's condition was so severe that it left him unable to work, care for himself, or live a normal life.

Born in Indonesia in 1971, Koswara lived a relatively normal life until he reached his teenage years. It was then that the first warts began to appear on his hands and feet. Over time, these growths became more extensive, covering his limbs and eventually resembling tree bark. The thick, bark-like lesions grew uncontrollably, making everyday tasks impossible and leading to severe physical and emotional suffering.

Epidermodysplasia verruciformis is an extremely rare condition caused by a genetic mutation that affects the body's ability to fight off certain types of human papillomavirus (HPV). In individuals with EV, HPV infections result in the formation of scaly macules and papules, which can develop into large, wart-like growths. These lesions have a characteristic appearance, resembling tree bark or wood. The condition is often referred to as "tree man syndrome" due to the striking and unusual appearance of the growths.

Koswara's case was particularly severe, with growths covering much of his body and significantly impacting his quality of life. Unable to work or perform basic tasks, he relied on his family for support. His condition also made him a target of social stigma and discrimination, further isolating him from his community.

In 2007, Koswara's story gained international attention when he was featured in a Discovery Channel documentary. The program highlighted his condition and the significant challenges he faced. Following the documentary, Koswara received medical attention from a team of doctors in Indonesia and the United States. The medical team, led by Indonesian dermatologist Dr. Rachmat Dinata and American dermatologist Dr. Anthony Gaspari, developed a treatment plan to help manage Koswara's condition.

Koswara underwent multiple surgeries to remove the massive growths on his hands and feet. Over a period of several years, doctors removed more than 6 kilograms (13 pounds) of warty tissue from his body. The surgeries provided some relief and improved his ability to perform daily activities. However, the growths continued to reoccur, requiring ongoing medical intervention.

Despite the challenges, Koswara's case has contributed to a greater understanding of epidermodysplasia verruciformis and its treatment. Researchers have studied his condition to better understand the genetic and viral factors involved in EV. This research has the potential to improve diagnosis, treatment, and management of the condition for other affected individuals.

Dede Koswara's story is a powerful reminder of the resilience of the human spirit in the face of adversity. His courage and determination to seek treatment and improve his quality of life have inspired many around the world. Although his condition was physically and emotionally challenging, Koswara's willingness to share his story has raised awareness about rare genetic disorders and the importance of medical research.

In 2016, after a long battle with his condition, Dede Koswara passed away due to complications related to his illness. His legacy lives on through the increased awareness and understanding of epidermodysplasia verruciformis that his story brought to the world. The "Tree Man" will be remembered not only for his unique condition but also for his bravery and the hope he inspired in others facing similar challenges.

4. The Man Who Can't Forget

Jill Price, often referred to as the woman with hyperthymesia, has an extraordinary and rare ability to remember virtually every detail of her life. Hyperthymesia, also known as highly superior autobiographical memory (HSAM), is a condition that allows individuals to recall events from their personal past with incredible accuracy and detail. Price's remarkable memory has provided valuable insights into the workings of human memory and has raised intriguing questions about the limits of human cognition.

Jill Price first came to public attention in 2006 when she contacted researchers at the University of California, Irvine, seeking help for her unusual memory. She described her ability to remember every day of her life since the age of 14 in vivid detail. This remarkable claim piqued the interest of neuroscientists Dr. James McGaugh and Dr. Larry Cahill, who began to study her memory abilities in depth.

Hyperthymesia is characterized by the ability to recall personal experiences and events with extraordinary detail and accuracy. Unlike individuals with photographic memory, who can recall visual information with precision, people with hyperthymesia have a deep and comprehensive recall of their autobiographical memories. They can remember dates, conversations, and even minute details about events that occurred decades ago.

For Jill Price, living with hyperthymesia has been both a gift and a curse. While her memory allows her to recall joyous moments with unparalleled clarity, it also means she cannot forget painful or traumatic experiences. This constant flood of memories can be overwhelming and emotionally exhausting. Price has described her condition as a "running movie" in her mind, with memories playing back continuously and involuntarily.

Scientific studies on Jill Price and other individuals with hyperthymesia have revealed fascinating insights into the mechanisms of human memory. Brain imaging studies have shown that people with hyperthymesia have larger and more active regions of the brain associated with autobiographical memory, such as the hippocampus and temporal lobes. These findings suggest that their exceptional memory abilities may be linked to structural and functional differences in their brains.

Researchers have also explored the cognitive and psychological aspects of hyperthymesia. Individuals with this condition often exhibit obsessive tendencies and spend a significant amount of time thinking about their past experiences. This constant reflection on personal memories may contribute to their ability to recall events with such detail and accuracy. However, it can also lead to difficulties in focusing on the present and planning for the future.

Jill Price's experiences have highlighted the complex interplay between memory and emotion. While her ability to remember every detail of her life can be advantageous, it also comes with significant emotional challenges. The inability to forget painful or distressing events can lead to heightened anxiety and depression.

Price has worked with therapists to develop coping strategies for managing her memories and maintaining her mental well-being.

The study of hyperthymesia has broader implications for our understanding of memory and cognition. It challenges the conventional view that forgetting is a flaw in the memory system. Instead, it suggests that forgetting may play a crucial role in maintaining mental health and cognitive function. The ability to forget irrelevant or distressing information allows individuals to focus on the present and adapt to new experiences.

Jill Price's story has inspired further research into the nature of memory and the potential for enhancing memory abilities. Scientists are exploring ways to harness the mechanisms underlying hyperthymesia to develop new treatments for memory-related disorders, such as Alzheimer's disease and other forms of dementia. Understanding how hyperthymesia works could lead to breakthroughs in preserving and enhancing memory function in the aging population.

In conclusion, Jill Price's extraordinary memory abilities have provided valuable insights into the complexities of human memory. Her experiences with hyperthymesia have highlighted the potential benefits and challenges of having an exceptional autobiographical memory. As researchers continue to study this rare condition, they hope to unlock the secrets of memory and find ways to enhance cognitive function for everyone. Jill Price's story serves as a reminder of the incredible capabilities of the human mind and the intricate balance between remembering and forgetting.

5. The Ice Man

Wim Hof, known as the "Ice Man," has gained worldwide fame for his extraordinary ability to withstand extreme cold. His feats include climbing Mount Everest and Mount Kilimanjaro in nothing but shorts, running a marathon in the Arctic Circle barefoot, and holding the world record for the longest ice bath. Hof attributes his abilities to a combination of breathing techniques, cold exposure, and meditation, which he collectively refers to as the Wim Hof Method.

Wim Hof was born on April 20, 1959, in Sittard, Netherlands. His journey into cold exposure began at a young age when he developed a fascination with the cold. This interest grew into a full-fledged practice, leading him to discover the physical and mental benefits of exposing his body to extreme cold temperatures. Hof's unique methods and seemingly superhuman abilities have attracted the attention of scientists and researchers, eager to understand the underlying mechanisms of his feats.

The Wim Hof Method consists of three primary components: breathing exercises, cold exposure, and commitment (or meditation). The breathing exercises involve a series of deep, rhythmic breaths followed by periods of breath-holding. This practice is designed to increase oxygen levels in the body, boost energy, and improve focus. Cold exposure, the second component, involves gradual and controlled exposure to cold environments, such as ice baths and cold showers. This practice is believed to strengthen the cardiovascular system, improve immune function, and increase mental resilience. The third component, commitment or meditation, emphasizes the importance of mental focus and determination in mastering the body's response to stress.

Hof's claims and abilities have been the subject of scientific scrutiny. Researchers have conducted several studies to investigate the physiological and psychological effects of the Wim Hof Method. One notable study, published in the Proceedings of the National Academy of Sciences, found that practitioners of the Wim Hof Method were able to voluntarily influence their autonomic nervous system and immune response. The study demonstrated that Hof's techniques could lead to increased production of anti-inflammatory proteins and a decrease in pro-inflammatory cytokines, suggesting potential benefits for conditions characterized by excessive inflammation.

Another study conducted by Radboud University in the Netherlands examined the effects of the Wim Hof Method on cold tolerance. Participants who practiced Hof's techniques showed a significant increase in their ability to tolerate cold temperatures, as well as improvements in their metabolic response to cold exposure. These findings support Hof's claims that his methods can enhance the body's ability to withstand extreme cold.

Wim Hof's feats have inspired many people around the world to adopt his methods and explore the limits of their own physical and mental capabilities. His teachings emphasize the power of the human mind and the body's ability to adapt to challenging conditions. Hof's approach has also gained popularity as a tool for improving overall health and well-being, with many practitioners reporting benefits such as increased energy, reduced stress, and enhanced mental clarity.

In addition to his work on cold exposure, Wim Hof has become an advocate for mental health awareness and stress management. He believes that his methods can help individuals cope with anxiety, depression, and other mental health challenges by promoting resilience and improving emotional regulation. Hof's own experiences with personal loss and adversity have shaped his mission to empower others and share the transformative potential of his techniques.

Wim Hof's story is a testament to the extraordinary potential of the human body and mind. His ability to endure extreme cold and his commitment to pushing the boundaries of human performance have made him a global phenomenon. As scientific research continues to explore the mechanisms behind Hof's methods, his contributions to the fields of health and wellness, mental resilience, and human potential remain profound and far-reaching.

In conclusion, Wim Hof, the "Ice Man," has captivated the world with his remarkable ability to withstand extreme cold and his innovative approach to health and wellness. Through his Wim Hof Method, he has demonstrated the power of the human mind and body to adapt and thrive under challenging conditions. Hof's journey from a curious child to a global icon of resilience and strength serves as an inspiration to all who seek to explore the limits of their own potential.

6. The Rubber Boy

Daniel Browning Smith, also known as "The Rubber Boy," has gained international fame for his extraordinary flexibility and contortionist abilities. Smith holds multiple Guinness World Records for his contortionist skills, making him one of the most flexible individuals in the world. His incredible talent has earned him appearances on numerous television shows, movies, and live performances, captivating audiences with his seemingly impossible body movements.

Born on May 8, 1979, in Meridian, Mississippi, Daniel Browning Smith discovered his talent for contortion at a young age. As a child, he was fascinated by circus performers and began practicing his flexibility skills. Smith's natural ability to bend and twist his body into unusual shapes quickly set him apart from his peers. His dedication to honing his craft led him to join a circus school, where he received formal training in contortion and acrobatics.

Smith's contortionist abilities are the result of a unique combination of genetic factors and intense training. He possesses hypermobile joints, a condition that allows for a greater range of motion in his joints than the average person. This condition, combined with his rigorous practice regimen, has enabled him to perform extraordinary feats of flexibility. Smith's training includes stretching exercises, strength training, and practicing contortionist routines to maintain and improve his skills.

Throughout his career, Daniel Browning Smith has achieved numerous accolades and broken multiple world records. Some of his most notable records include the fastest time to travel 20 meters in a backbend position, the most dislocations of a shoulder within one minute, and the most flexible man in the world. His ability to contort his body into seemingly impossible positions has earned him a place in the Guinness World Records and recognition as one of the top contortionists globally.

In addition to his world records, Smith's contortionist abilities have led to a successful career in the entertainment industry. He has appeared on various television shows, including "America's Got Talent," "The Tonight Show with Jay Leno," and "Stan Lee's Superhumans." Smith has also performed in movies, live shows, and circuses worldwide, showcasing his unique talent to diverse audiences. His performances often leave spectators in awe, as he bends, twists, and contorts his body in ways that defy normal human capabilities.

Beyond his entertainment career, Daniel Browning Smith's contortionist abilities have also contributed to scientific research. Researchers have studied his body to understand the biological mechanisms behind his extraordinary flexibility. These studies have provided insights into joint hypermobility, connective tissue disorders, and the potential for enhancing flexibility through training. Smith's participation in scientific research has helped to advance knowledge in the fields of anatomy, physiology, and biomechanics.

Smith's journey as a contortionist has not been without challenges. The physical demands of contortion can lead to injuries, including joint dislocations, muscle strains, and ligament tears. To mitigate these risks, Smith follows a strict regimen of conditioning exercises, stretching, and strength training to maintain his body's resilience and prevent injuries. His commitment to his craft and his ability to manage the physical challenges of contortion have been key to his long-lasting success.

Daniel Browning Smith's story is a testament to the incredible potential of the human body and the power of dedication and perseverance. His ability to push the limits of flexibility and contortion has inspired countless individuals to pursue their own unique talents and passions. Smith's achievements highlight the importance of recognizing and nurturing exceptional abilities, regardless of how unconventional they may seem.

In conclusion, Daniel Browning Smith, "The Rubber Boy," has captivated the world with his extraordinary flexibility and contortionist skills. His journey from a curious child to a world-renowned contortionist showcases the incredible capabilities of the human body and the power of dedication and training. Smith's contributions to the fields of entertainment and scientific research have left a lasting impact, inspiring others to explore the limits of their own potential and celebrate the unique talents that make us all extraordinary.

7.   The Man with the Iron Stomach

Michel Lotito, known as "Monsieur Mangetout" (Mr. Eat-All), was a French entertainer famous for his ability to consume and digest unusual and indigestible objects. Born on June 15, 1950, in Grenoble, France, Lotito amazed the world with his unique talent for eating metal, glass, rubber, and other materials that would be harmful or impossible for an average person to digest. His extraordinary ability earned him a place in the Guinness World Records and made him a global sensation.

111

Lotito's unusual eating habits began at a young age. At nine years old, he discovered that he enjoyed eating glass and metal. Over time, he developed a remarkable tolerance for consuming such materials, and by adulthood, he had turned his peculiar talent into a career as a professional entertainer. Lotito's performances involved him eating items like bicycles, televisions, shopping carts, and even a Cessna 150 airplane, which he consumed piece by piece over two years from 1978 to 1980.

The scientific community was intrigued by Lotito's ability to consume and digest indigestible objects without suffering harm. Medical examinations revealed that Lotito had a thickened stomach lining and intestines, which allowed him to ingest sharp and hazardous materials without injury. His digestive system also produced unusually powerful gastric juices, enabling him to break down and process the materials he consumed. These unique physiological traits distinguished Lotito from the average person and allowed him to pursue his extraordinary diet.

To facilitate the consumption of metal and other hard objects, Lotito would break them into small pieces and drink significant amounts of mineral oil and water to help his digestive system process the materials. Despite the apparent risks, Lotito claimed that he rarely experienced any discomfort or health issues related to his unusual diet. He continued to eat indigestible objects throughout his life, performing in various countries and captivating audiences with his remarkable ability.

Lotito's performances were not only a testament to his unique physiological traits but also to his mental fortitude and determination. Eating objects such as glass and metal requires a high level of mental discipline and control to overcome the natural aversion to consuming dangerous materials. Lotito's ability to do so consistently and safely highlighted his exceptional mental and physical resilience.

112

Michel Lotito's legacy extends beyond his remarkable talent for eating indigestible objects. His story has contributed to a better understanding of the limits and capabilities of the human body. Researchers have studied Lotito's physiological adaptations to gain insights into the digestive system and the body's ability to cope with extreme conditions. His case has also sparked interest in exploring the genetic and environmental factors that can lead to such extraordinary abilities.

Lotito's achievements were recognized globally, and he received numerous accolades and honors for his unique talent. In 2007, he was awarded the Guinness World Record for the "strangest diet" after consuming an estimated nine tons of metal over his lifetime. His ability to eat and digest objects that would be harmful to most people earned him a place in history as one of the most unusual and fascinating individuals.

Michel Lotito passed away on June 25, 2007, at the age of 57. Despite his unusual diet, his death was not related to his eating habits but was caused by natural causes. Lotito's legacy lives on through the stories of his incredible feats and the scientific insights gained from studying his unique abilities. His life serves as a reminder of the extraordinary potential of the human body and the fascinating diversity of human capabilities.

In conclusion, Michel Lotito, "Monsieur Mangetout," astounded the world with his ability to consume and digest indigestible objects. His unique physiological traits and mental resilience enabled him to perform feats that seemed impossible to most people. Lotito's story highlights the incredible adaptability of the human body and the limitless potential of human abilities. His legacy continues to inspire curiosity and wonder about the extraordinary capacities of the human body and mind.

8.   The Human Pincushion

Tim Cridland, also known by his stage name "Zamora the Torture King," has captivated audiences worldwide with his extraordinary ability to withstand extreme pain and perform seemingly impossible acts of endurance. Cridland is famous for his acts of self-inflicted torture, including piercing his body with skewers, lying on beds of nails, and walking on hot coals. His remarkable tolerance for pain and his unique performances have earned him a place among the most unusual entertainers in the world.

Tim Cridland was born on May 3, 1963, in Tacoma, Washington. From an early age, he was fascinated by the limits of human endurance and the power of the mind over the body. This curiosity led him to explore various techniques of pain management and body control. As a teenager, Cridland began experimenting with piercing his skin and other acts of self-inflicted pain. Over time, he developed a remarkable ability to control his body's response to pain, allowing him to perform acts that would be unbearable for most people.

Cridland's unique talent for withstanding pain is the result of both psychological and physiological factors. He has trained himself to enter a state of deep concentration and control, allowing him to manage his body's reaction to pain. This mental discipline, combined with his unique physical tolerance, enables him to perform his extraordinary feats. Some researchers believe that Cridland's ability to withstand pain may be linked to a genetic predisposition that allows his nervous system to process pain differently from most people.

One of Cridland's most famous acts is piercing his body with long skewers, often through his arms, cheeks, and neck. These piercings are done in a way that avoids major blood vessels and vital organs, but the visual impact on audiences is profound. Cridland has also performed acts such as lying on beds of nails, walking on broken

114

glass, and subjecting himself to electric shocks. Each performance showcases his incredible pain tolerance and his ability to control his body's response to extreme stimuli.

Cridland's performances have earned him a reputation as one of the most extreme entertainers in the world. He has appeared on numerous television shows, including "Ripley's Believe It or Not!," "Stan Lee's Superhumans," and "The Guinness World Records Show." His acts have fascinated and horrified audiences, drawing both admiration and curiosity about the limits of human endurance.

In addition to his performances, Tim Cridland has also contributed to scientific research on pain tolerance and the human nervous system. Researchers have studied his body and brain to understand the mechanisms behind his extraordinary abilities. These studies have provided insights into pain management techniques, the psychology of pain, and the potential for enhancing pain tolerance through training and mental conditioning.

Cridland's journey as a performer has not been without risks. The physical demands of his acts can lead to injuries and complications. However, Cridland takes great care to ensure his safety and minimize the risks associated with his performances. He follows strict protocols for piercing and other acts of self-inflicted pain, and he regularly consults with medical professionals to maintain his health and well-being.

Beyond his performances, Tim Cridland's story is one of perseverance and the pursuit of human potential. His ability to withstand extreme pain and push the boundaries of human endurance serves as an inspiration to those who seek to explore the limits of their own capabilities. Cridland's journey highlights the power of the mind over the body and the incredible potential of human resilience.

In conclusion, Tim Cridland, "Zamora the Torture King," has captivated audiences with his extraordinary ability to withstand extreme pain and perform acts of self-inflicted torture. His unique talent is the result of both psychological and physiological factors, allowing him to control his body's response to pain in ways that most people cannot. Cridland's story is a testament to the incredible potential of the human mind and body, inspiring curiosity and wonder about the limits of human endurance and resilience.

9. The Stone Man Syndrome (Fibrodysplasia Ossificans Progressiva)

Fibrodysplasia Ossificans Progressiva (FOP), often referred to as Stone Man Syndrome, is one of the most debilitating and rare genetic disorders known to science. This condition causes soft tissues, such as muscles, tendons, and ligaments, to gradually turn into bone, effectively imprisoning the body in an extra skeleton. The condition is so rare that it affects approximately 1 in 2 million people worldwide.

The characteristics and effects of Stone Man Syndrome are as dramatic as they are devastating:

Symptoms and Manifestations: The hallmark of FOP is the progressive transformation of soft tissues into bone, leading to restricted movement and eventual immobility. This process often begins in early childhood, with the first signs being malformations of the big toes, which are present at birth. As the individual ages, episodes of painful inflammation, known as flare-ups, cause the soft tissues to ossify (turn into bone). These flare-ups can be spontaneous or triggered by trauma, including minor injuries or surgical procedures.

116

Diagnosis: Diagnosing FOP is primarily clinical and based on the presence of characteristic symptoms such as toe malformations and progressive heterotopic ossification (abnormal bone formation in soft tissues). Genetic testing can confirm the diagnosis by identifying a mutation in the ACVR1 gene, which is responsible for the condition. Early and accurate diagnosis is crucial to prevent unnecessary surgeries, which can exacerbate the condition.

Genetic Causes: FOP is caused by a mutation in the ACVR1 gene, which encodes a receptor involved in bone growth and development. The mutation leads to the receptor being overactive, causing inappropriate bone formation. This genetic defect is typically sporadic, meaning it usually occurs as a new mutation in the affected individual and is not inherited from the parents.

Management and Treatment: Currently, there is no cure for FOP, and treatment focuses on managing symptoms and preventing flare-ups. Avoiding injuries and minimizing trauma are essential to reducing the risk of new bone formation. Medications such as corticosteroids may help manage acute flare-ups, while muscle relaxants and pain relievers can provide symptomatic relief. Physical therapy is generally limited due to the risk of inducing flare-ups. Research is ongoing to find treatments that can slow or stop the progression of the disease, with gene therapy and targeted drugs being explored as potential options.

Psychological and Social Impacts: Living with FOP poses significant psychological and social challenges. Individuals with the condition often face severe physical limitations, chronic pain, and dependency on caregivers. The progressive nature of the disease can lead to feelings of isolation, anxiety, and depression. Support from healthcare professionals, family, friends, and patient organizations is crucial in helping affected individuals maintain their quality of life and mental well-being.

117

Notable Cases: One of the most well-documented cases of FOP is that of Harry Eastlack, whose skeleton, after his death, was donated to medical research and is now displayed at the Mütter Museum in Philadelphia. His case has provided valuable insights into the progression of the disease and continues to aid researchers in their quest to understand and treat FOP.

Research and Future Directions: Research into FOP is focused on understanding the molecular mechanisms underlying the condition and developing targeted therapies. Advances in genetic research and biotechnology hold promise for the future. For instance, scientists are exploring the use of small molecules to inhibit the abnormal activity of the ACVR1 receptor, and gene editing technologies like CRISPR to correct the genetic mutation. Patient registries and collaborative research efforts are also playing a crucial role in advancing our understanding of FOP and accelerating the development of new treatments.

Fibrodysplasia Ossificans Progressiva, or Stone Man Syndrome, remains one of the most challenging genetic conditions faced by those who are affected. Despite the severe limitations it imposes, ongoing research and the resilience of those living with the condition provide hope for future advancements in treatment and care. Understanding and addressing the complexities of FOP continues to be a critical area of focus in medical research, with the goal of improving the lives of those affected by this relentless disease.

10. The Werewolf Syndrome (Hypertrichosis)

Hypertrichosis, often referred to as Werewolf Syndrome, is an extremely rare condition characterized by excessive hair growth on the body, including areas where hair usually does not grow. This condition can be congenital (present at birth) or acquired later in life, and it affects both men and women. The rarity and dramatic appearance of hypertrichosis have made it a subject of fascination and myth throughout history.

The characteristics and causes of Hypertrichosis are as intriguing as they are diverse:

Symptoms and Manifestations: The primary symptom of hypertrichosis is the abnormal growth of hair, which can be fine, thick, short, or long. This excessive hair growth can occur on the face, limbs, torso, and even palms and soles. In congenital cases, hair growth is often evident from birth or early childhood, while acquired hypertrichosis can develop at any age and is often linked to other medical conditions or the use of certain medications.

Types of Hypertrichosis:

Congenital Hypertrichosis Lanuginosa: This form is characterized by the presence of lanugo, the fine hair that normally covers a fetus and is shed before birth. In individuals with this condition, the lanugo persists into adulthood.

Congenital Generalized Hypertrichosis: Individuals with this form have dense hair growth over most of their body, including areas typically free of hair.

Acquired Hypertrichosis: This type can develop as a result of medical conditions such as anorexia nervosa, certain cancers, or side effects from medications like minoxidil or cyclosporine.

Genetic Causes: Congenital hypertrichosis is often inherited and linked to genetic mutations. These mutations can lead to the activation of hair growth in areas where hair follicles are usually inactive. The exact genes involved vary, and research is ongoing to identify all the genetic factors contributing to the condition.

Acquired Causes: Acquired hypertrichosis can be triggered by a variety of factors, including hormonal imbalances, metabolic disorders, and the use of certain drugs. For instance, prolonged use of steroids or anti-seizure medications has been associated with increased hair growth.

119

Historical and Cultural Context: Throughout history, individuals with hypertrichosis have often been subjects of curiosity and spectacle. During the 19th and early 20th centuries, people with this condition were frequently exhibited in circuses and sideshows as "werewolves" or "wolf men." These exhibitions contributed to the mythologizing of the condition and often led to social stigma and isolation for those affected.

Management and Treatment: There is no cure for hypertrichosis, but various treatments can help manage the condition. Hair removal methods such as shaving, waxing, laser therapy, and electrolysis can reduce the visibility of excessive hair. In cases where hypertrichosis is linked to medication or an underlying medical condition, addressing the primary cause can help alleviate the symptoms. Genetic counseling may be beneficial for individuals with congenital forms of the condition.

Psychosocial Impact: Living with hypertrichosis can be challenging due to the social and psychological implications of the condition. Individuals may experience social stigma, discrimination, and emotional distress related to their appearance. Psychological support, including counseling and support groups, can help individuals cope with these challenges and improve their quality of life.

Research and Future Directions: Ongoing research aims to better understand the genetic and molecular mechanisms underlying hypertrichosis. Advances in genetic research and dermatology may lead to more effective treatments and interventions in the future. Public awareness and education about hypertrichosis can also help reduce stigma and promote a more inclusive and understanding society.

Hypertrichosis, or Werewolf Syndrome, highlights the incredible diversity of human biology and the complex interplay between genetics, environment, and health. While the condition poses significant challenges, the resilience and strength of those living with hypertrichosis inspire ongoing efforts to understand and manage this rare anomaly. Greater awareness and scientific advancements hold the promise of improved treatments and a better quality.

Chapter 6: Astonishing Animal Abilities

1. The Mantis Shrimp's Super Vision

The mantis shrimp, a small marine crustacean, possesses one of the most extraordinary visual systems in the animal kingdom. Found in tropical and subtropical waters, these colorful creatures are known for their powerful claws and exceptional eyesight. The mantis shrimp's vision is so advanced that it can detect polarized light and see a wider range of colors than any other known animal, including ultraviolet and infrared light.

Mantis shrimps belong to the order Stomatopoda and are divided into two main groups: spearers and smashers. Despite their small size, typically ranging from 10 to 20 centimeters, mantis shrimps are formidable predators. Their vision plays a crucial role in hunting, navigation, and communication within their complex underwater environment.

The secret behind the mantis shrimp's super vision lies in its highly specialized eyes. Unlike humans, who have three types of color-receptive cones (red, green, and blue), mantis shrimps have up to sixteen types of photoreceptor cells. This vast array of photoreceptors allows them to perceive an incredible spectrum of colors. Moreover, their eyes can move independently, giving them a wide field of view and the ability to track multiple objects simultaneously.

One of the most remarkable aspects of mantis shrimp vision is their ability to detect polarized light. Polarization refers to the orientation of light waves, and while many animals can sense polarized light to some extent, mantis shrimps can see it in much finer detail. This ability is thought to be used for enhancing contrast in murky waters, detecting prey, and communicating with other mantis shrimps through body coloration that changes under polarized light.

The mantis shrimp's eyes are divided into three distinct parts: the dorsal and ventral hemispheres, and a central band called the midband. The midband contains specialized ommatidia (the optical units of compound eyes) responsible for color and polarized light detection. Each of the midband rows is tuned to different wavelengths of light, allowing the mantis shrimp to break down the light spectrum into more components than human eyes can.

Researchers have been fascinated by the mantis shrimp's vision for decades. Studies have shown that the complex visual system of these creatures offers potential applications in technology. For instance, cameras and imaging devices inspired by mantis shrimp eyes could be used to improve underwater exploration, medical imaging, and even the detection of cancer cells, which can exhibit polarized light properties.

The mantis shrimp's visual abilities are not only scientifically intriguing but also essential for its survival. In addition to hunting and navigating, mantis shrimps use their vision to communicate with potential mates and rivals. Their vibrant colors and patterns, which can appear differently under polarized light, convey important information about their identity, reproductive status, and aggression levels.

In the wild, mantis shrimps exhibit a variety of behaviors that showcase their extraordinary vision. They can accurately strike at prey with their powerful claws, often faster than a blink of an eye, thanks to their ability to judge distances and detect subtle movements. Their vision also helps them avoid predators and navigate complex reef environments.

Despite their formidable abilities, mantis shrimps face threats from habitat destruction and climate change. Coral reefs, which provide essential shelter and hunting grounds for mantis shrimps, are under threat from rising ocean temperatures and acidification. Conservation efforts aimed at protecting these delicate ecosystems are crucial for the survival of mantis shrimps and countless other marine species.

In conclusion, the mantis shrimp's super vision is a marvel of natural evolution. With their ability to see a vast range of colors and detect polarized light, these small but mighty creatures have one of the most advanced visual systems in the animal kingdom. Their remarkable eyesight not only aids in their survival but also offers valuable insights and potential applications for human technology. The mantis shrimp stands as a testament to the incredible diversity and complexity of life on Earth.

2. The Mimic Octopus

The mimic octopus (Thaumoctopus mimicus) is a master of disguise, renowned for its extraordinary ability to imitate a wide variety of marine animals. Discovered in 1998 off the coast of Indonesia, this remarkable cephalopod has captivated scientists and nature enthusiasts with its uncanny mimicry skills. The mimic octopus can transform its appearance to resemble other creatures such as lionfish, flatfish, and sea snakes, making it one of the most versatile and fascinating inhabitants of the ocean.

The mimic octopus's habitat consists primarily of the sandy and muddy seafloor of shallow coastal waters in the Indo-Pacific region. This environment, with its relatively sparse hiding places, makes the octopus's mimicry abilities crucial for avoiding predators and hunting prey. The octopus achieves its incredible transformations through a combination of rapid color changes, body posturing, and movement patterns that convincingly replicate the appearance and behavior of other marine animals.

The primary mechanism behind the mimic octopus's transformations lies in its skin. Like other octopuses, it has specialized skin cells called chromatophores, which contain pigments that can expand or contract to change the color of the skin. Additionally, iridophores and leucophores reflect and scatter light, further enhancing the color and texture changes. By controlling these cells, the mimic octopus can quickly alter its appearance to blend in with its surroundings or take on the guise of a different species.

What sets the mimic octopus apart from other cephalopods is its ability to replicate the shapes and movements of multiple species. For example, when threatened by predators such as damselfish, the mimic octopus can assume the appearance of a banded sea snake by extending two arms and undulating them to mimic the snake's movements. This tactic deters potential attackers that would typically avoid the venomous sea snake.

In addition to the sea snake, the mimic octopus can imitate the highly venomous lionfish by spreading its arms wide and displaying bold stripes. It can also flatten its body and glide along the seafloor to resemble a flatfish, such as a sole or flounder. These disguises serve not only to avoid predators but also to approach prey unsuspectingly, showcasing the octopus's cunning and adaptability.

The mimic octopus's mimicry abilities have intrigued scientists and prompted extensive research into the evolutionary advantages and mechanisms behind such remarkable adaptations. Studies have shown that the mimic octopus's behavior likely evolved as a response to the high predation pressure in its habitat. By imitating dangerous or unpalatable species, the octopus increases its chances of survival and reduces the likelihood of predation.

Researchers are also interested in the neural and cognitive processes that enable the mimic octopus to perform such complex imitations. The octopus's nervous system is highly sophisticated, with a large portion of its neurons located in its arms. This decentralized nervous system allows for precise control over arm movements and skin changes, enabling the rapid and accurate mimicking of other species.

The mimic octopus's ability to imitate multiple species highlights the incredible versatility and intelligence of cephalopods. It challenges our understanding of animal behavior and cognition, demonstrating that even relatively small and short-lived creatures can exhibit complex and adaptive behaviors.

In conclusion, the mimic octopus is a true marvel of the marine world. Its extraordinary ability to transform its appearance and behavior to mimic other species serves as a powerful defense mechanism and an effective hunting strategy. The mimic octopus's remarkable mimicry skills underscore the incredible diversity and adaptability of life in the ocean and continue to inspire scientific curiosity and wonder.

3. The Immortal Jellyfish

The Turritopsis dohrnii, commonly known as the "immortal jellyfish," has captivated scientists and the public alike with its unique ability to revert to its juvenile form after reaching maturity. This remarkable capability allows the jellyfish to effectively bypass death, making it a subject of intense interest in the fields of biology and aging research. Found in the Mediterranean Sea and the waters off Japan, the immortal jellyfish is a small, translucent creature that holds the key to understanding the mysteries of aging and regeneration.

The life cycle of the immortal jellyfish begins like that of many other jellyfish species. It starts as a tiny larva, known as a planula, which settles on the seafloor and develops into a polyp. The polyp then gives rise to multiple juvenile jellyfish, or medusae, through a process called budding. Unlike other jellyfish, however, the Turritopsis dohrnii has the extraordinary ability to reverse its life cycle. When faced with environmental stress, physical damage, or even the natural aging process, the jellyfish can transform its cells back into a polyp stage, effectively starting its life cycle anew.

This process of reverting to an earlier developmental stage is known as transdifferentiation. During transdifferentiation, the cells of the jellyfish undergo a transformation, changing from one type to another. For example, muscle cells can become nerve cells, and reproductive cells can become skin cells. This cellular plasticity allows the jellyfish to regenerate its entire body and return to a youthful state.

The biological mechanisms underlying transdifferentiation in the immortal jellyfish are of great interest to scientists, particularly in the context of aging and regenerative medicine. Researchers are studying the jellyfish's genome and cellular processes to identify the genes and molecular pathways involved in its remarkable regenerative abilities. Understanding these mechanisms could have profound implications for human health, potentially leading to advances in tissue regeneration, wound healing, and even treatments for age-related diseases.

The potential applications of studying the immortal jellyfish extend beyond aging research. The insights gained from this tiny creature could also inform the development of new medical therapies and biotechnological innovations. For example, the jellyfish's ability to

126

reverse cellular aging could inspire techniques for preserving and rejuvenating human cells, improving organ transplantation outcomes, and extending the lifespan of biomedical tissues.

Despite its extraordinary abilities, the immortal jellyfish faces challenges in its natural environment. Predation, disease, and changing ocean conditions all pose threats to its survival. However, its unique regenerative capabilities offer a significant advantage, allowing it to recover from injuries and adverse conditions that would be fatal to other species.

In conclusion, the Turritopsis dohrnii, or immortal jellyfish, is a fascinating organism that defies the conventional understanding of aging and mortality. Its ability to revert to a juvenile form through transdifferentiation opens up exciting possibilities for scientific research and medical advancements. By studying the immortal jellyfish, scientists hope to unlock the secrets of cellular regeneration and aging, potentially paving the way for breakthroughs in human health and longevity.

4. The Electric Eel

The electric eel (Electrophorus electricus) is a remarkable freshwater fish known for its ability to generate powerful electric shocks. Native to the rivers and swamps of South America, the electric eel uses its electric organs for hunting, self-defense, and navigation. This unique adaptation has made the electric eel a subject of fascination for scientists and a symbol of the incredible diversity of life in the Amazon Basin.

The electric eel's ability to produce electricity is facilitated by specialized cells called electrocytes, which are derived from muscle cells. These electrocytes are stacked in series within three main electric organs: the main organ, the Hunter's organ, and the Sach's organ. When the eel needs to generate an electric shock, it simultaneously activates the electrocytes, creating a rapid discharge of electrical energy. This process can produce shocks of up to 600 volts, enough to stun prey and deter potential predators.

The electric eel uses its electric discharges for various purposes. During hunting, the eel emits high-voltage pulses to stun or kill prey, such as fish, amphibians, and invertebrates. The electric shock incapacitates the prey, making it easier for the eel to capture and consume it. Additionally, the electric eel uses lower-voltage discharges for navigation and communication. These weaker electric signals help the eel locate objects in its environment, detect prey, and communicate with other electric eels.

The electric eel's ability to generate electricity has intrigued scientists for centuries. Early researchers, including the famous naturalist Alexander von Humboldt, documented the electric eel's shocking capabilities during expeditions to South America. Modern scientists continue to study the eel's electric organs to understand the underlying biological and biophysical mechanisms.

The potential applications of electric eel research are vast. By studying the eel's electrocytes and electric organs, scientists hope to develop new bio-inspired technologies. For example, research on electric eels has contributed to the development of bio-batteries and other energy storage devices. Understanding how the eel generates and controls electrical discharges could also inform the design of new medical devices, such as pacemakers and neural stimulators.

In addition to its electric abilities, the electric eel exhibits fascinating behaviors and adaptations. It can breathe air, allowing it to survive in oxygen-poor waters. The eel's elongated, snake-like body is well-suited for navigating the complex underwater environments of the Amazon Basin. Its ability to generate electric shocks provides a significant advantage in the dense and murky waters where visibility is often limited.

Despite its name, the electric eel is not a true eel but a member of the knifefish family. Its classification as a knifefish highlights the diverse evolutionary paths that can lead to similar adaptations in different groups of animals.

In conclusion, the electric eel is a remarkable example of nature's ingenuity. Its ability to generate powerful electric shocks for hunting, defense, and navigation sets it apart as one of the most fascinating creatures in the animal kingdom. The electric eel's unique adaptations continue to inspire scientific research and technological innovation, showcasing the incredible potential of bio-inspired design and the importance of preserving biodiversity in the Amazon Basin.

**Should we modify this page or shall we move on to the next page? Tell me what to modify or if you feel it's good as it is....Shall

5. The Lyrebird's Mimicry

The lyrebird, native to Australia, is renowned for its extraordinary ability to mimic natural and artificial sounds with incredible accuracy. There are two species of lyrebird: the superb lyrebird (Menura novaehollandiae) and Albert's lyrebird (Menura alberti). Both species are celebrated for their intricate vocalizations and elaborate courtship displays. The lyrebird's mimicry skills make it one of the most remarkable songbirds in the world, showcasing the incredible adaptability and intelligence of avian species.

Lyrebirds are named for their striking tail feathers, which resemble the shape of a lyre, a classical musical instrument. During the breeding season, male lyrebirds use their impressive tails and vocal abilities to attract females. They perform elaborate displays, spreading their tails and singing a complex medley of sounds. These sounds can include the songs of other birds, the calls of animals, and even human-made noises such as chainsaws, car alarms, and camera shutters.

The secret to the lyrebird's mimicry lies in its highly developed syrinx, the vocal organ of birds. The syrinx allows lyrebirds to produce a wide range of sounds by controlling the airflow through their vocal cords with exceptional precision. This enables them to replicate sounds with remarkable accuracy and detail. Lyrebirds spend years learning and perfecting their repertoire, often incorporating the sounds they hear in their environment into their songs.

The primary purpose of the lyrebird's mimicry is to attract mates. During the breeding season, males establish and defend territories, where they perform their vocal displays to entice females. The complexity and diversity of a male's song can indicate his fitness and genetic quality, making him more attractive to potential mates. In addition to attracting females, the lyrebird's mimicry can also serve to deter rivals and predators by imitating the calls of more aggressive species.

The superb lyrebird is particularly famous for its mimicry abilities. It has been observed imitating the songs of more than 20 different bird species, as well as various other natural and artificial sounds. The lyrebird's ability to adapt its song to its environment demonstrates its incredible auditory memory and learning capacity. This adaptability has allowed lyrebirds to thrive in diverse habitats, from rainforests to eucalypt woodlands.

Albert's lyrebird, while less studied than the superb lyrebird, also exhibits impressive mimicry skills. Found in a limited range in southeastern Queensland, Albert's lyrebird is known for its powerful and melodious song, which includes imitations of other birds and environmental sounds. The conservation of Albert's lyrebird is of particular concern due to its restricted habitat and the threats posed by habitat destruction and climate change.

The lyrebird's mimicry abilities have fascinated ornithologists and bird enthusiasts for centuries. Early European settlers in Australia documented the lyrebird's remarkable vocal talents, and it continues to be a subject of scientific research and public interest. Researchers study the lyrebird's vocalizations to understand the mechanisms of avian mimicry, the evolution of complex communication systems, and the cognitive abilities of birds.

The lyrebird's mimicry also highlights the broader importance of sound in the natural world. Many animals rely on vocalizations for communication, navigation, and survival. The lyrebird's ability to imitate a wide range of sounds underscores the significance of acoustic environments and the need to preserve natural habitats that support diverse soundscapes.

In conclusion, the lyrebird's extraordinary mimicry abilities make it one of the most remarkable songbirds in the world. Its capacity to replicate a wide array of sounds with astonishing accuracy showcases the adaptability and intelligence of avian species. The lyrebird's vocal talents play a crucial role in its courtship displays and highlight the importance of sound in animal communication. The study and conservation of lyrebirds continue to inspire awe and wonder, emphasizing the incredible diversity of life on Earth.

6. The Archerfish's Precision Shooting

The archerfish, native to the mangroves and estuaries of Southeast Asia and Australia, is known for its remarkable ability to shoot jets of water to knock down insects and other prey from overhanging vegetation. This unique hunting strategy sets the archerfish apart as one of nature's most skilled marksmen, showcasing the incredible precision and adaptability of aquatic life.

The archerfish (Toxotes jaculatrix and related species) uses its specialized mouth and tongue to form a tube, allowing it to shoot a jet of water with impressive accuracy and force. This behavior is particularly fascinating because it requires the fish to compensate for the refraction of light at the water's surface. Refraction bends light rays as they pass from air to water, causing objects above the surface to appear displaced from their actual positions. Despite this optical challenge, archerfish consistently hit their targets with remarkable precision.

Researchers have studied the archerfish's shooting technique to understand how these fish achieve such accuracy. Studies have shown that archerfish are able to learn and improve their aim over time through trial and error. They use visual cues and adjust the angle and force of their water jets based on their distance from the target. This ability to learn and adapt demonstrates a level of cognitive flexibility that is rare among fish.

The archerfish's shooting mechanism involves rapid and coordinated movements of the mouth and tongue. When the fish spots a target, it positions itself just below the water's surface, aligns its body, and quickly snaps its mouth and tongue to create a jet of water. The force of the jet is strong enough to dislodge insects and other small prey from leaves and branches, causing them to fall into the water where the archerfish can easily capture them.

In addition to shooting water jets, archerfish have also been observed leaping out of the water to snatch prey directly from vegetation. This versatility in hunting techniques further highlights their adaptability and problem-solving abilities. Archerfish are capable of switching between different strategies based on the specific circumstances of their environment and the behavior of their prey.

The study of archerfish has provided valuable insights into animal behavior and cognition. Researchers have discovered that archerfish can use observational learning, where they watch and learn from the actions of other fish. This ability to learn from others is a significant cognitive skill that is more commonly associated with mammals and birds than with fish. It suggests that archerfish possess a higher level of social intelligence than previously thought.

The archerfish's unique hunting strategy also has potential applications in technology and engineering. Understanding the principles behind their precise water jets could inspire innovations in fluid dynamics and the design of water-based propulsion systems. Additionally, the archerfish's ability to adjust for optical refraction could inform the development of advanced targeting systems and optical devices.

Despite their impressive abilities, archerfish face challenges in their natural habitats. Habitat destruction, pollution, and climate change threaten the mangroves and estuaries they depend on for survival. Conservation efforts aimed at protecting these vital ecosystems are essential for ensuring the continued existence of archerfish and other species that rely on these environments.

In conclusion, the archerfish's precision shooting is a remarkable example of nature's ingenuity. Their ability to shoot jets of water with incredible accuracy, compensate for optical refraction, and adapt their hunting strategies showcases the incredible adaptability and cognitive abilities of aquatic life. The archerfish's unique skills continue to fascinate scientists and inspire a deeper appreciation for the complexity and diversity of life in our planet's waters.

7.   The Pistol Shrimp's Sonic Weapon

The pistol shrimp, also known as the snapping shrimp, is a small but formidable marine creature known for its powerful claw that can create one of the loudest sounds in the ocean. With its unique ability to generate a high-velocity water jet and a loud snap, the pistol shrimp uses this sonic weapon to stun prey and defend itself from predators. This remarkable adaptation showcases the incredible diversity of survival strategies in the animal kingdom.

Pistol shrimp belong to the family Alpheidae, and there are several species found in warm coastal waters around the world. These shrimp are typically small, ranging from 3 to 5 centimeters in length, but they possess an oversized claw that can be nearly half the length of their body. This claw is not only a physical weapon but also a sophisticated tool for creating powerful underwater shockwaves.

The mechanism behind the pistol shrimp's snap is both fascinating and complex. The shrimp's claw has a specialized joint that allows it to open and close rapidly. When the claw is closed quickly, it shoots out a high-speed jet of water. This action creates a cavitation bubble—a pocket of low pressure that forms in the water. When the bubble collapses, it produces a loud snapping sound and a shockwave that can reach up to 218 decibels, a noise level comparable to a gunshot. The implosion of the bubble generates a brief flash of light and temperatures as high as the surface of the sun, making it a powerful and versatile weapon.

The pistol shrimp uses its sonic weapon primarily for hunting and defense. When hunting, the shrimp lies in wait for small fish, crabs, or other prey to pass by. It then snaps its claw, creating a shockwave that can stun or kill the prey instantly. The shrimp quickly grabs the immobilized prey with its smaller claw and brings it to its burrow to consume. This hunting technique is highly effective and allows the pistol shrimp to capture prey much larger than itself.

In addition to hunting, the pistol shrimp uses its snapping claw for communication and territorial defense. The loud snap serves as a warning signal to other shrimp and potential predators, indicating the presence of a powerful and well-armed defender. The sound can deter predators and competitors, helping the shrimp maintain control over its territory.

The symbiotic relationship between some species of pistol shrimp and gobies is another fascinating aspect of their behavior. In this mutualistic relationship, the goby fish and the shrimp share a burrow, with the shrimp maintaining and defending the burrow and the goby providing protection by acting as a lookout. The goby uses its excellent vision to detect approaching predators and warns the shrimp by flicking its tail. In return, the shrimp allows the goby to share its burrow, providing a safe refuge. This partnership showcases the intricate and cooperative behaviors that can evolve in the animal kingdom.

The pistol shrimp's sonic abilities have attracted the interest of researchers and engineers. Scientists study the shrimp's snapping mechanism to understand the physics of cavitation and the potential applications of this phenomenon. Research on pistol shrimp has informed the development of new underwater communication and navigation technologies, as well as materials and devices designed to withstand extreme conditions.

In conclusion, the pistol shrimp's powerful snapping claw and its ability to generate high-velocity water jets and loud shockwaves make it one of the most remarkable marine creatures. This unique adaptation highlights the incredible diversity of survival strategies in the animal kingdom and the complex interactions between different species. The pistol shrimp's sonic weapon continues to inspire scientific research and technological innovation, showcasing the extraordinary potential of nature's ingenuity.

8. The Leafcutter Ant's Agricultural Skills

The leafcutter ant, primarily found in the tropical forests of Central and South America, is known for its extraordinary agricultural abilities. These ants, belonging to the genera Atta and Acromyrmex, have developed one of the most advanced and efficient farming systems in the animal kingdom. Leafcutter ants

135

harvest leaves not for direct consumption but to cultivate a specialized fungus, which serves as their primary food source. This symbiotic relationship between the ants and the fungus showcases the remarkable complexity of insect societies and their ability to engage in sophisticated agricultural practices.

Leafcutter ants live in large, complex colonies that can contain millions of individuals. The colony is organized into a highly structured caste system, with different ants performing specific roles. The primary castes include the queen, who is responsible for laying eggs; workers, who gather leaves and tend to the fungus gardens; soldiers, who protect the colony; and minor workers, who care for the queen and larvae. This division of labor ensures the efficient functioning and survival of the colony.

The process of leaf harvesting begins with worker ants venturing out of the nest in long lines, climbing trees and cutting leaves with their powerful mandibles. Despite the large size of the leaf fragments relative to their body size, the ants carry the pieces back to the nest, forming impressive trails of leaf-bearing ants. Once the leaves reach the nest, they are processed by smaller workers who chew them into a pulp. This pulp serves as a substrate for the cultivation of the fungus, Leucoagaricus gongylophorus.

The relationship between leafcutter ants and their cultivated fungus is a prime example of mutualism, where both species benefit from the association. The fungus relies on the ants to provide it with a constant supply of leaf substrate, while the ants depend on the fungus for nourishment. The ants carefully tend to the fungus, maintaining optimal conditions for its growth by regulating temperature, humidity, and removing any contaminants. They also secrete antimicrobial substances to protect the fungus from harmful pathogens.

One of the most fascinating aspects of leafcutter ant agriculture is their ability to manage their fungal crops with remarkable precision.

The ants are capable of recognizing and removing any fungal strains that are not beneficial or are infected by parasites. They also have a unique way of distributing the fungus throughout the colony, ensuring that all members have access to the nutritious food source.

Research on leafcutter ants has provided valuable insights into the evolution of agriculture and the complex social behaviors of insects. Studies have shown that the ants' farming practices have been evolving for millions of years, making them one of the earliest known agriculturalists. The efficiency and sustainability of their farming methods offer potential lessons for human agricultural practices, particularly in areas such as pest management and sustainable cultivation.

Leafcutter ants also play a crucial role in their ecosystems. By harvesting leaves, they influence plant composition and diversity in their habitats. Their activities contribute to nutrient cycling, soil aeration, and the overall health of the forest ecosystem. The leafcutter ants' farming practices demonstrate the intricate connections between species and the importance of biodiversity in maintaining ecological balance.

Despite their impressive abilities, leafcutter ants face threats from habitat destruction and climate change. Deforestation and land conversion for agriculture reduce the availability of suitable habitats for these ants, while changing climate conditions can disrupt their delicate fungal cultivation systems. Conservation efforts aimed at preserving tropical forests and their biodiversity are essential for ensuring the survival of leafcutter ants and their unique agricultural systems.

In conclusion, the leafcutter ant's agricultural skills are a testament to the incredible adaptability and ingenuity of insect societies. Their sophisticated farming practices, mutualistic relationship with fungus, and complex social organization highlight the remarkable

137

capabilities of these tiny creatures. The study of leafcutter ants continues to provide valuable insights into the evolution of agriculture, the dynamics of mutualistic relationships, and the importance of biodiversity in maintaining healthy ecosystems.

9. The Cuttlefish's Camouflage Mastery

Cuttlefish, belonging to the class Cephalopoda, are remarkable marine animals known for their extraordinary ability to change color and texture in an instant. This ability allows cuttlefish to blend seamlessly into their surroundings, communicate with other cuttlefish, and deter predators. Found in oceans around the world, cuttlefish are often referred to as the "chameleons of the sea" due to their impressive camouflage skills.

The cuttlefish's camouflage abilities are primarily driven by specialized cells in their skin called chromatophores, iridophores, and leucophores. Chromatophores contain pigments that can be expanded or contracted to change the animal's color. These cells are controlled by the nervous system, allowing for rapid color changes. Iridophores reflect light, creating iridescent colors, while leucophores scatter light, adding to the overall camouflage effect.

The cuttlefish's skin is incredibly versatile, allowing it to mimic various textures and patterns in its environment. This texture-changing ability is due to muscular control over small structures called papillae on its skin. By adjusting the shape and size of these papillae, cuttlefish can create a range of textures, from smooth to spiky, enhancing their ability to blend into different backgrounds.

Cuttlefish use their camouflage for several purposes, the most important of which is predator avoidance. By matching their surroundings, they can effectively hide from predators such as sharks and larger fish. This ability to disappear into the background is crucial for their survival, especially in the complex and diverse habitats of coral reefs and rocky seabeds.

In addition to avoiding predators, cuttlefish use their color-changing abilities for hunting. As ambush predators, they often lie in wait for prey to come within reach. By camouflaging themselves, they can remain hidden until the perfect moment to strike. Cuttlefish primarily feed on small fish, crabs, and other crustaceans, using their rapid color changes to confuse and capture their prey.

Cuttlefish also use their camouflage skills for communication and mating displays. During courtship, males display vibrant patterns to attract females and assert dominance over other males. These displays can include rapid changes in color and intricate patterns that convey specific messages. The ability to change color and pattern quickly allows cuttlefish to communicate complex signals in a dynamic and visually rich underwater environment.

The study of cuttlefish camouflage has provided valuable insights into the neural and muscular control of color and texture changes. Researchers are particularly interested in how cuttlefish perceive their environment and how their brains process visual information to control their skin cells. Understanding these mechanisms could have applications in developing advanced materials and technologies, such as adaptive camouflage for military use and dynamic displays for electronic devices.

Despite their impressive abilities, cuttlefish face challenges in their natural habitats. Overfishing, habitat destruction, and climate change pose significant threats to their populations. Conservation efforts are essential to protect these fascinating creatures and the ecosystems they inhabit.

In conclusion, cuttlefish are masters of camouflage, capable of changing their color and texture with remarkable speed and precision. Their ability to blend into their surroundings, communicate through visual displays, and hunt with stealth showcases the incredible adaptability and intelligence of these marine animals. The study of cuttlefish continues to inspire scientific research and technological innovation, highlighting the remarkable diversity and complexity of life in the ocean.

10. The Honeybee's Dance Language

Honeybees, primarily the species Apis mellifera, are not only known for their vital role in pollination and honey production but also for their remarkable method of communication known as the "waggle dance." This intricate dance language allows honeybees to convey detailed information about the location of food sources to their hive mates. The waggle dance is a testament to the complex social behaviors and communication systems that have evolved in insect societies.

The discovery of the waggle dance was made by Austrian ethologist Karl von Frisch in the 1940s. His groundbreaking research demonstrated that honeybees use specific movements to communicate the distance and direction of a food source relative to the hive. This form of communication is crucial for the efficiency and survival of the colony, as it enables worker bees to forage effectively and bring back resources to support the hive.

The waggle dance takes place on the vertical surfaces of the honeycomb within the hive. When a foraging bee discovers a rich source of nectar or pollen, she returns to the hive and performs a dance that consists of a series of movements, including a figure-eight pattern and a "waggle run." The waggle run is the most critical part of the dance, as it conveys the direction and distance to the food source.

During the waggle run, the bee moves in a straight line while waggling her abdomen from side to side. The angle of the waggle run relative to the vertical axis of the hive indicates the direction of the food source concerning the position of the sun. For example, if the bee waggles at a 60-degree angle to the right of the vertical, the food source is located 60 degrees to the right of the sun's current position. The duration of the waggle run indicates the distance to the food source; longer waggle runs correspond to greater distances.

In addition to direction and distance, the waggle dance can also convey information about the quality of the food source. The vigor and repetition of the dance can signal the abundance and richness of the nectar or pollen. This detailed communication allows other foraging bees to make informed decisions about where to collect resources, optimizing the colony's foraging efficiency.

The waggle dance is not the only form of communication used by honeybees. They also perform a "round dance" to indicate the presence of food sources located close to the hive, typically within 50 meters. The round dance consists of circular movements without the waggle run, signaling nearby foragers to search in the immediate vicinity for food.

The study of honeybee communication has provided valuable insights into animal behavior, navigation, and social organization. Researchers have used advanced techniques, such as high-speed video analysis and harmonic radar tracking, to observe and analyze the details of the waggle dance. These studies have revealed the precision and accuracy of the dance language, as well as the bees' ability to adjust their communication based on environmental factors such as wind and light conditions.

The honeybee's dance language has also inspired research in robotics and artificial intelligence. By understanding the principles of bee communication and navigation, scientists have developed algorithms and systems for swarm robotics, where multiple robots work together to accomplish tasks. The efficiency and coordination demonstrated by honeybee colonies serve as a model for designing autonomous robotic systems.

Despite their remarkable abilities, honeybees face numerous threats, including habitat loss, pesticide exposure, disease, and climate change. The decline in honeybee populations has significant implications for global agriculture and ecosystems, as bees are essential pollinators for many crops and wild plants. Conservation efforts aimed at protecting honeybee habitats and reducing the impact of harmful practices are crucial for ensuring the survival of these vital pollinators.

In conclusion, the honeybee's waggle dance is a sophisticated form of communication that exemplifies the complexity and intelligence of insect societies. This dance language enables honeybees to convey precise information about food sources, enhancing the colony's foraging efficiency and survival. The study of honeybee communication continues to inspire scientific research and technological innovation, highlighting the extraordinary capabilities of these small but essential creatures.

Should we modify this page or shall we move on to the next

Chapter 7: Incredible Natural Phenomena

1.   The Aurora Borealis

The Aurora Borealis, commonly known as the Northern Lights, is one of the most mesmerizing natural phenomena visible in the polar regions. This spectacular light show, characterized by shimmering waves of green, red, yellow, blue, and violet, has fascinated humans for centuries. The Aurora Borealis is primarily observed in high-latitude regions around the Arctic, but its southern counterpart, the Aurora Australis, occurs around the Antarctic.

The origins of the Aurora Borealis lie in the interactions between the solar wind and Earth's magnetosphere. The solar wind is a stream of charged particles, mainly electrons and protons, emitted by the sun. When these particles reach Earth, they are guided by the planet's magnetic field towards the polar regions. As the charged particles collide with the gases in Earth's upper atmosphere, primarily oxygen and nitrogen, they release energy in the form of light. The different colors of the aurora are determined by the type of gas involved in the collision and the altitude at which the interaction occurs.

Oxygen atoms at higher altitudes (above 150 miles) produce red auroras, while at lower altitudes (up to 150 miles), they emit green light, which is the most common color seen in the aurora. Nitrogen molecules, on the other hand, can produce blue or purplish-red auroras depending on the energy of the collision. The combination of these colors can result in a dynamic and multi-hued display that dances across the night sky.

The best time to observe the Aurora Borealis is during the winter months, from September to March, when the nights are longest and the skies are darkest. Areas with little to no light pollution, such as remote parts of Alaska, Canada, Norway, Sweden, Finland, and Iceland, provide the most favorable conditions for viewing this natural spectacle. However, strong geomagnetic storms can sometimes make the aurora visible at lower latitudes, offering a rare treat for observers in more temperate regions.

The aurora has inspired countless myths, legends, and scientific inquiries throughout history. Indigenous cultures in the Arctic, such as the Inuit and Sámi, have long held beliefs and stories about the Northern Lights. Some saw them as the spirits of ancestors, while others interpreted them as omens or messages from the gods.

143

In medieval Europe, the aurora was often seen as a sign of impending doom or divine intervention. Scientific understanding of the Aurora Borealis began to develop in the early 17th century when the phenomenon was first studied systematically. The term "aurora borealis" was coined by the Italian astronomer Galileo Galilei in 1619, combining the name of the Roman goddess of dawn, Aurora, with the Greek name for the north wind, Boreas. However, it was not until the 19th and 20th centuries that scientists fully understood the role of the solar wind and Earth's magnetic field in creating the aurora.

Modern technology has allowed scientists to study the Aurora Borealis in greater detail. Satellites and ground-based observatories equipped with sensitive instruments have provided valuable data on the solar wind, geomagnetic activity, and the upper atmosphere. These studies have not only enhanced our understanding of the aurora but also contributed to the field of space weather research. Space weather, which includes phenomena such as solar flares and geomagnetic storms, can have significant impacts on satellite communications, navigation systems, and power grids on Earth.

The Aurora Borealis is not only a subject of scientific interest but also a major attraction for tourists. Each year, thousands of people travel to polar regions hoping to witness the enchanting light display. In response, many countries have developed tourism industries centered around aurora viewing, offering guided tours, accommodations, and educational programs to enhance the experience for visitors.

In conclusion, the Aurora Borealis is a captivating natural phenomenon that showcases the beauty and complexity of our planet's interactions with the sun. Its vibrant colors and dynamic movements have inspired awe and wonder in people throughout history. As scientific research continues to uncover the mysteries of the aurora, this magnificent light show remains a powerful reminder of the intricate and interconnected nature of our universe.

2. The Great Blue Hole

The Great Blue Hole, located off the coast of Belize, is one of the most famous and visually stunning underwater sinkholes in the world. This giant marine cavern, circular in shape and surrounded by a ring of coral, measures about 318 meters (1,043 feet) in diameter and 124 meters (407 feet) in depth. It is part of the larger Belize Barrier Reef Reserve System, a UNESCO World Heritage site, and attracts divers and marine enthusiasts from around the globe.

The origins of the Great Blue Hole date back to the last Ice Age, over 150,000 years ago. During this period, sea levels were much lower than they are today, and the area that is now the Blue Hole was a dry limestone cave system. As the Ice Age ended and sea levels rose, the caves were flooded, and the roof of the cavern eventually collapsed, creating the massive sinkhole we see today. The result is a near-perfect circular formation that plunges deep into the Caribbean Sea.

One of the most intriguing aspects of the Great Blue Hole is its striking color contrast. The deep, rich blue of the sinkhole's waters stands in stark contrast to the lighter blue and turquoise of the surrounding shallow waters and coral reefs. This vivid color difference is due to the depth of the hole and the way light interacts with the water. The deeper waters absorb more light, resulting in a darker blue hue.

The Great Blue Hole is renowned for its unique geological formations and rich marine life. Divers who explore its depths can encounter stunning stalactites and stalagmites, remnants of the cave's ancient past. These formations, some of which are over 40

145

feet long, provide a glimpse into the geological history of the area and the processes that shaped it.

The marine ecosystem within and around the Great Blue Hole is diverse and vibrant. The surrounding coral reef is home to a variety of species, including parrotfish, angelfish, butterflyfish, and several species of sharks, such as Caribbean reef sharks, bull sharks, and hammerhead sharks. The deep waters of the Blue Hole also provide a habitat for more elusive and rare marine creatures, making it a fascinating destination for marine biologists and divers alike.

The Great Blue Hole was popularized by the legendary marine explorer Jacques Cousteau, who visited the site in 1971. Cousteau declared it one of the top ten scuba diving sites in the world, bringing international attention to the Blue Hole and Belize's barrier reef system. His exploration and documentation of the site helped to highlight its ecological and geological significance.

Diving in the Great Blue Hole is considered an advanced activity due to its depth and the potential challenges posed by underwater currents and limited visibility at greater depths. Divers typically descend along the walls of the sinkhole, exploring the stalactites and stalagmites at around 40 meters (130 feet) below the surface. The experience offers a unique combination of natural beauty, adventure, and a sense of exploring an ancient geological formation.

The Great Blue Hole is not only a natural wonder but also a valuable site for scientific research. Studies of the Blue Hole's geological formations have provided insights into past climate conditions, sea level changes, and the natural history of the Caribbean region. Additionally, research on the marine life within the Blue Hole contributes to our understanding of marine biodiversity and the health of coral reef ecosystems.

Despite its beauty and significance, the Great Blue Hole and the surrounding Belize Barrier Reef face environmental threats from climate change, overfishing, and pollution. Rising sea temperatures and ocean acidification pose significant risks to coral reefs worldwide, including those in Belize. Conservation efforts are essential to protect these fragile ecosystems and ensure that future generations can continue to marvel at the Great Blue Hole's natural splendor.

In conclusion, the Great Blue Hole of Belize is a breathtaking natural phenomenon that showcases the incredible geological and marine diversity of our planet. Its origins as an ancient limestone cave, coupled with its vibrant marine ecosystem and striking visual appeal, make it a must-see destination for divers and nature enthusiasts. As we continue to study and protect this unique site, the Great Blue Hole stands as a testament to the awe-inspiring power and beauty of the natural world.

3. The Catatumbo Lightning

The Catatumbo Lightning is a mesmerizing and powerful natural phenomenon that occurs over the Catatumbo River in Venezuela. This remarkable display of lightning storms, also known as the "everlasting storm," is characterized by frequent and intense lightning flashes that illuminate the night sky. The Catatumbo Lightning is unique due to its consistency and intensity, making it one of the most extraordinary weather phenomena in the world.

The Catatumbo Lightning occurs primarily over the marshlands where the Catatumbo River empties into Lake Maracaibo. This region, known for its humid and warm climate, creates the perfect conditions for the development of thunderstorms. The phenomenon typically occurs between April and November, with the highest frequency of lightning strikes happening during the rainy season. On average, the Catatumbo Lightning can produce lightning storms on 140 to 160 nights per year, with up to 280 lightning strikes per hour.

The scientific explanation behind the Catatumbo Lightning lies in the unique topography and atmospheric conditions of the region. The warm, moist air from the Caribbean Sea collides with the cooler air descending from the Andes Mountains, creating strong updrafts that fuel the formation of thunderstorms. Additionally, the presence of methane from the marshlands may contribute to the intensity of the lightning strikes. These combined factors result in the consistent and powerful lightning storms that characterize the Catatumbo Lightning.

The lightning storms of Catatumbo have significant cultural and historical importance in Venezuela. For centuries, the indigenous people of the region, such as the Wayuu, have observed and revered the Catatumbo Lightning. The phenomenon has also played a role in Venezuelan folklore and mythology, often seen as a sign of protection and power. In modern times, the Catatumbo Lightning has become a symbol of Venezuelan natural heritage and a point of national pride.

The Catatumbo Lightning is not only a visual spectacle but also an important source of atmospheric ozone. The lightning strikes contribute to the production of ozone by breaking down nitrogen and oxygen molecules in the atmosphere, leading to the formation of ozone. This process helps to replenish the ozone layer, which is crucial for protecting life on Earth from harmful ultraviolet radiation. However, the contribution of the Catatumbo Lightning to the global ozone balance is relatively small compared to other natural and human-made sources.

Researchers and scientists are fascinated by the Catatumbo Lightning and have conducted various studies to understand its mechanisms and impacts. These studies involve monitoring the frequency, intensity, and distribution of lightning strikes, as well as analyzing the atmospheric conditions that give rise to the phenomenon. Advanced technologies, such as satellite imagery and lightning detection networks, have provided valuable data for understanding the dynamics of the Catatumbo Lightning.

The Catatumbo Lightning also attracts tourists and photographers from around the world who seek to witness and capture the breathtaking display of nature's power. Guided tours to the region offer visitors the opportunity to experience the lightning storms up close, providing a unique and unforgettable adventure. Efforts to promote sustainable tourism in the area aim to balance the economic benefits of tourism with the need to protect the fragile ecosystems of the Catatumbo River basin.

Despite its awe-inspiring beauty, the Catatumbo Lightning and the surrounding region face environmental challenges. Deforestation, pollution, and climate change pose threats to the delicate balance of the ecosystem that supports this natural phenomenon. Conservation efforts are essential to preserve the integrity of the Catatumbo River basin and ensure the continued occurrence of the Catatumbo Lightning for future generations to marvel at.

In conclusion, the Catatumbo Lightning is a spectacular natural phenomenon that showcases the incredible power and beauty of our planet's weather systems. Its unique and consistent lightning storms over the Catatumbo River in Venezuela have fascinated and inspired people for centuries. As scientific research continues to unravel the mysteries of the Catatumbo Lightning, this extraordinary display of nature remains a testament to the dynamic and interconnected forces that shape our world.

4.   The Great Barrier Reef

The Great Barrier Reef, located off the northeastern coast of Australia, is the largest coral reef system in the world. Stretching over 2,300 kilometers (1,430 miles) and covering an area of approximately 344,400 square kilometers (133,000 square miles), it is composed of nearly 3,000 individual reefs and 900 islands. The Great Barrier Reef is renowned for its breathtaking beauty, incredible biodiversity, and ecological significance, making it one of the most remarkable natural phenomena on Earth.

The formation of the Great Barrier Reef began millions of years ago, with the accumulation of coral polyps—tiny, soft-bodied organisms that build hard, calcium carbonate exoskeletons. Over time, these coral polyps multiply and form colonies, creating the complex and diverse structures that make up the reef. The warm, shallow waters of the Coral Sea provide ideal conditions for coral growth, allowing the reef to flourish and support a wide variety of marine life.

The biodiversity of the Great Barrier Reef is unparalleled, hosting thousands of species of marine organisms. It is home to over 1,500 species of fish, 400 species of coral, 4,000 species of mollusks, and countless other marine creatures, including sea turtles, sharks, rays, and marine mammals. The reef's vibrant and colorful coral formations provide critical habitats and breeding grounds for many of these species, contributing to the overall health and productivity of the ocean ecosystem.

The ecological significance of the Great Barrier Reef extends beyond its immediate vicinity. The reef acts as a natural barrier, protecting the Australian coastline from the erosive forces of waves and storms. It also plays a crucial role in nutrient cycling and carbon sequestration, helping to regulate the Earth's climate and maintain the health of the marine environment. The intricate relationships between the reef's inhabitants create a delicate balance that sustains the diverse and dynamic ecosystem.

In addition to its ecological importance, the Great Barrier Reef is a major tourist destination, attracting millions of visitors each year. Tourists from around the world come to experience the reef's

150

stunning underwater landscapes through activities such as snorkeling, scuba diving, and boat tours. The tourism industry provides significant economic benefits to the region, supporting local communities and businesses. Efforts to promote sustainable tourism aim to ensure that the reef's natural beauty and resources are preserved for future generations.

Despite its remarkable resilience, the Great Barrier Reef faces numerous threats that jeopardize its survival. Climate change is one of the most significant challenges, as rising sea temperatures and ocean acidification can lead to coral bleaching and the degradation of reef structures. Coral bleaching occurs when corals expel the symbiotic algae that provide them with energy and color, leaving them vulnerable to disease and death. Mass bleaching events have become more frequent and severe in recent years, raising concerns about the long-term health of the reef.

Other threats to the Great Barrier Reef include pollution, overfishing, coastal development, and invasive species. Runoff from agricultural and industrial activities can introduce harmful chemicals and sediments into the reef's waters, disrupting the delicate balance of the ecosystem. Overfishing and destructive fishing practices can deplete key species and damage coral habitats, while coastal development can lead to habitat loss and increased sedimentation. Invasive species, such as the crown-of-thorns starfish, can also pose significant risks to coral health by preying on the coral polyps.

Conservation efforts are essential to protect and restore the Great Barrier Reef. These efforts include measures to reduce carbon emissions, improve water quality, regulate fishing practices, and manage coastal development. Scientific research and monitoring programs are critical for understanding the reef's health and developing effective conservation strategies. Public awareness and education campaigns also play a vital role in promoting responsible behaviors and fostering a sense of stewardship for this natural wonder.

5.   The Salar de Uyuni

The Salar de Uyuni, located in southwest Bolivia, is the world's largest salt flat, covering an area of over 10,000 square kilometers (3,900 square miles). This extraordinary landscape, created by the evaporation of ancient lakes, is a dazzling expanse of bright white salt crust that stretches as far as the eye can see. The Salar de Uyuni is not only a stunning natural wonder but also a significant resource for lithium and other minerals, making it an important site both ecologically and economically.

The formation of the Salar de Uyuni began over 30,000 years ago with the existence of prehistoric lakes, including Lake Minchin. Over thousands of years, these lakes dried up due to climate changes, leaving behind vast deposits of salt and other minerals. The result is a flat, reflective surface that creates a mirror-like effect when covered with a thin layer of water, especially during the rainy season from November to March. This phenomenon transforms the salt flat into a surreal landscape where the sky and ground seem to merge into one.

The Salar de Uyuni is composed primarily of sodium chloride (table salt), along with other minerals such as lithium, potassium, and magnesium. Beneath the thick salt crust lies a brine reservoir rich in lithium, a key component in rechargeable batteries for electronics and electric vehicles. Bolivia's lithium reserves in the Salar de Uyuni are among the largest in the world, making it a vital resource for the global transition to renewable energy and sustainable technologies.

In addition to its mineral wealth, the Salar de Uyuni is home to unique ecosystems and a variety of wildlife. During the rainy season, the salt flat becomes a breeding ground for flamingos,

152

including the Andean, Chilean, and James's flamingos. These birds feed on the brine shrimp and algae that thrive in the salty waters. The surrounding high-altitude landscape, known as the Altiplano, is also home to other species such as vicuñas, llamas, and various cacti and hardy vegetation adapted to the harsh environment.

The Salar de Uyuni is a popular tourist destination, attracting visitors from around the world who come to witness its otherworldly beauty and unique features. The vast, open space provides a perfect backdrop for photography, especially during the mirror effect in the rainy season. Tourists often take creative perspective photos that play with the flat, featureless landscape. Additionally, the nearby town of Uyuni serves as a base for tours that explore the salt flat and the surrounding areas, including the colorful Laguna Colorada and the eerie train cemetery.

The salt flat's unique optical properties have also made it a valuable tool for satellite calibration. The flat, featureless surface provides an ideal reference point for calibrating satellite instruments and ensuring accurate measurements of the Earth's surface. This application underscores the scientific importance of the Salar de Uyuni beyond its ecological and economic significance.

Despite its many attractions and resources, the Salar de Uyuni faces environmental challenges. The extraction of lithium and other minerals must be managed carefully to avoid damaging the delicate ecosystem and depleting the region's natural resources. Responsible mining practices, environmental regulations, and sustainable tourism are essential to preserving the Salar de Uyuni's unique landscape and biodiversity.

In conclusion, the Salar de Uyuni is a breathtaking natural wonder that exemplifies the beauty and diversity of our planet's landscapes. Its vast expanse of salt, mirror-like reflections, and rich mineral deposits make it a site of ecological, economic, and scientific significance. As we continue to explore and utilize the resources of

153

the Salar de Uyuni, it is crucial to balance development with conservation efforts to ensure that this extraordinary natural phenomenon remains intact for future generations to admire and study.

6. The Grand Prismatic Spring

The Grand Prismatic Spring, located in Yellowstone National Park, Wyoming, is the largest hot spring in the United States and the third largest in the world. This stunning natural feature is renowned for its vibrant colors, which create a striking, rainbow-like effect that can be seen from above. The Grand Prismatic Spring is a testament to the geothermal activity that shapes Yellowstone's unique landscape and supports a diverse array of microbial life.

The Grand Prismatic Spring measures approximately 110 meters (360 feet) in diameter and is about 50 meters (160 feet) deep. The vivid colors that characterize the spring are produced by the different types of heat-loving bacteria that thrive in the varying temperatures of the water. The center of the spring is a deep blue, created by the purity and depth of the water, which absorbs all wavelengths of light except blue. Surrounding the blue center are bands of green, yellow, orange, and red, created by thermophilic (heat-loving) bacteria and archaea that live in the cooler, mineral-rich waters around the edges of the spring.

The colors of the Grand Prismatic Spring change with the seasons. In the summer, the outer edges of the spring are dominated by orange and red hues, while in the winter, cooler temperatures result in a more muted palette of greens and blues. This seasonal variation is due to changes in the microbial populations and the pigments they produce, which adapt to the varying environmental conditions.

The heat that powers the Grand Prismatic Spring comes from the Yellowstone Caldera, an enormous volcanic system beneath the park. The caldera was formed by a series of massive volcanic eruptions, the last of which occurred approximately 640,000 years ago. Today, geothermal activity continues to shape the landscape of Yellowstone, producing hot springs, geysers, fumaroles, and mud pots. The heat from the caldera warms the groundwater, which then rises to the surface, creating the spectacular hot springs for which Yellowstone is famous.

The Grand Prismatic Spring is not only a visual marvel but also a site of significant scientific interest. Researchers study the unique microbial communities that inhabit the spring to gain insights into the origins of life on Earth and the potential for life on other planets. The extremophiles that thrive in the harsh conditions of the spring provide valuable information about the adaptability and resilience of life. These studies have implications for fields such as astrobiology, biotechnology, and environmental science.

Visiting the Grand Prismatic Spring is a highlight for many Yellowstone National Park visitors. The best way to view the spring is from the nearby overlook on the Fairy Falls Trail, which offers a panoramic view of the vibrant colors and patterns. Visitors can also walk along the boardwalk that runs alongside the spring, providing close-up views of the hot water and microbial mats. It is important for visitors to stay on designated paths and boardwalks to protect the delicate microbial communities and ensure their own safety, as the hot water can cause severe burns.

The preservation of the Grand Prismatic Spring and Yellowstone's other geothermal features is a priority for park management. Efforts to protect the park's natural resources include monitoring geothermal activity, regulating visitor access, and educating the public about the importance of conservation. These measures help to ensure that future generations can continue to enjoy and study the unique geothermal wonders of Yellowstone.

In conclusion, the Grand Prismatic Spring is a spectacular example of the natural beauty and geothermal activity that define Yellowstone National Park. Its vibrant colors, shaped by heat-loving microorganisms, create a stunning visual display that attracts visitors from around the world. As a site of scientific research, the Grand Prismatic Spring offers valuable insights into the origins and adaptability of life. Conservation efforts are essential to preserve this extraordinary natural phenomenon for future generations to appreciate and study.

7. The Bioluminescent Bays of Puerto Rico

The bioluminescent bays of Puerto Rico are among the most magical and unique natural phenomena in the world. These bays, where the water glows with a blue-green light when disturbed, are caused by the presence of bioluminescent microorganisms called dinoflagellates. The most famous bioluminescent bays in Puerto Rico are Mosquito Bay on the island of Vieques, Laguna Grande in Fajardo, and La Parguera in Lajas. Each offers visitors a chance to witness a natural light show that is both awe-inspiring and scientifically fascinating.

Bioluminescence is the production and emission of light by living organisms. In the bioluminescent bays of Puerto Rico, the primary culprits are dinoflagellates, specifically the species Pyrodinium bahamense. These tiny, single-celled organisms emit light as a chemical reaction when they are agitated, creating the stunning glow seen in the water. When the water is disturbed by movement, such as a hand swirling in the water or a paddle dipping in, the dinoflagellates emit light, producing a shimmering effect.

The ideal conditions for bioluminescence include warm, nutrient-rich waters with low circulation, which allow the dinoflagellates to thrive. Puerto Rico's bays provide these perfect conditions, with their sheltered environments and abundant organic material. The

156

concentration of dinoflagellates in these bays can reach up to hundreds of thousands per gallon of water, creating a dense population that enhances the brightness of the bioluminescence. Mosquito Bay on Vieques is often considered the brightest bioluminescent bay in the world. It has earned this reputation due to its high concentration of dinoflagellates and its relatively undisturbed natural environment. Visitors to Mosquito Bay can experience the bioluminescence by kayaking or taking boat tours, especially on moonless nights when the glow is most visible. The sight of the water lighting up with each stroke of the paddle or splash is an unforgettable experience that leaves a lasting impression.

Laguna Grande in Fajardo and La Parguera in Lajas also offer incredible bioluminescent experiences. Laguna Grande is unique for its narrow canal that connects the lagoon to the ocean, creating a dramatic effect as boats pass through and disturb the water. La Parguera is known for its more accessible location and the variety of tour options available, including snorkeling and swimming in the bioluminescent waters.

The phenomenon of bioluminescence has intrigued scientists and researchers for decades. Studying bioluminescent organisms helps scientists understand the mechanisms of light production in nature and its various functions, such as predator avoidance, communication, and attracting mates. The enzyme luciferase and the substrate luciferin are key components in the biochemical reactions that produce light in dinoflagellates and other bioluminescent organisms.

Beyond their scientific interest, the bioluminescent bays of Puerto Rico are significant for their ecological and cultural value. The presence of these glowing waters highlights the importance of preserving natural habitats and maintaining the health of marine ecosystems. Efforts to protect the bays include regulating tourism, reducing pollution, and promoting sustainable practices to ensure that these natural wonders can be enjoyed by future generations. The bioluminescent bays also hold cultural significance for the local communities. They are a source of pride and a major attraction that supports the local economy through tourism. Educational programs and guided tours help raise awareness about the importance of conservation and the unique marine life that inhabits these waters.

However, the bioluminescent bays face several threats, including climate change, pollution, and overuse from tourism. Rising temperatures and changes in water chemistry can affect the delicate balance needed for dinoflagellates to thrive. Pollution from agricultural runoff, sewage, and coastal development can reduce water quality and harm the ecosystems that support bioluminescence. Sustainable tourism practices and environmental regulations are crucial for protecting these fragile environments.

In conclusion, the bioluminescent bays of Puerto Rico offer a spectacular natural light show that captivates visitors and underscores the wonder of our planet's marine life. The glowing waters, produced by bioluminescent dinoflagellates, create an enchanting experience that combines beauty, science, and environmental importance. Preserving these natural phenomena requires continued efforts in conservation and sustainable practices to ensure that the magic of the bioluminescent bays can be appreciated for generations to come.

8.   The Great Red Spot of Jupiter

The Great Red Spot on Jupiter is one of the most iconic and enduring features in our solar system. This gigantic storm, larger than Earth itself, has been raging for at least 350 years and possibly much longer. The Great Red Spot is a massive anticyclonic storm, characterized by its reddish color and swirling cloud patterns. It serves as a testament to the dynamic and turbulent atmosphere of the largest planet in our solar system.

The Great Red Spot measures about 16,350 kilometers (10,159 miles) in width, although its size has fluctuated over time. It was once much larger, spanning nearly three times the diameter of Earth. Observations since the 19th century have documented a gradual shrinking of the storm, but it remains an immense and powerful feature. The storm rotates counterclockwise, completing a full rotation approximately every six days.

The reddish hue of the Great Red Spot is one of its most distinctive characteristics, though the exact cause of the color is still a matter of scientific investigation. Theories suggest that the color may be due to chemical compounds in Jupiter's atmosphere, such as complex organic molecules, red phosphorus, or compounds involving sulfur. The intense radiation and ultraviolet light from the Sun interacting with these chemicals may contribute to the storm's red appearance.

The Great Red Spot is a high-pressure region in Jupiter's atmosphere, with winds reaching speeds of up to 432 kilometers per hour (268 miles per hour). This contrasts with hurricanes on Earth, which are low-pressure systems. The vertical structure of the storm extends deep into Jupiter's atmosphere, though its exact depth is still not fully understood. Observations from spacecraft such as NASA's Juno mission have provided valuable data on the storm's dynamics and structure.

The persistence of the Great Red Spot is another intriguing aspect. While storms on Earth typically last for days to weeks, the Great Red Spot has endured for centuries. Its longevity is likely due to Jupiter's lack of a solid surface, which means there are no landforms to disrupt the storm's circulation. Additionally, the planet's rapid rotation and immense size contribute to the stability and energy of the storm.

The Great Red Spot has been a focus of scientific study since its discovery. Early observations date back to the 17th century, with Italian astronomer Giovanni Cassini often credited with the first recorded sighting. Since then, it has been continuously monitored by telescopes on Earth and various space missions, including the Pioneer, Voyager, Galileo, and Juno missions.

The study of the Great Red Spot provides valuable insights into atmospheric dynamics and planetary science. By understanding the mechanisms driving this enormous storm, scientists can draw parallels to atmospheric processes on Earth and other planets. The data collected from studying Jupiter's atmosphere helps improve our understanding of fluid dynamics, weather patterns, and the behavior of giant gas planets.

In addition to its scientific significance, the Great Red Spot captures the imagination and curiosity of people around the world. Its colossal size and enduring presence make it a symbol of the power and beauty of our solar system's natural phenomena. The storm is a popular subject for astronomy enthusiasts and educators, often featured in textbooks, documentaries, and planetarium shows.

The Great Red Spot's future remains uncertain. While it has shown signs of shrinking, it continues to be a dominant feature on Jupiter. Ongoing observations and future missions to Jupiter will help scientists monitor changes in the storm and further unravel the mysteries of this gigantic atmospheric phenomenon.

In conclusion, the Great Red Spot of Jupiter is a remarkable and enduring natural phenomenon that showcases the dynamic and turbulent nature of the largest planet in our solar system. Its immense size, distinctive color, and long-lasting presence make it a subject of fascination and scientific inquiry. As we continue to explore and study Jupiter, the Great Red Spot remains a powerful reminder of the complexity and grandeur of the universe.

Should we modify this page or shall we move on to the next page? Tell me what to modify or if you feel it's good as it is.

9.   The Eye of the Sahara

The Eye of the Sahara, also known as the Richat Structure, is a fascinating and mysterious geological formation located in the Sahara Desert in Mauritania, Africa. This large, circular feature resembles a giant bullseye and is easily visible from space, spanning approximately 40 kilometers (25 miles) in diameter. The Eye of the Sahara has intrigued scientists and geologists for decades, leading to various theories about its formation and significance.

The Richat Structure was first identified in the 1960s during space missions, when astronauts noted its striking appearance. Initially thought to be an impact crater from a meteorite, further studies revealed that the formation is not the result of an extraterrestrial collision. Instead, the Eye of the Sahara is now understood to be a deeply eroded geological dome, created by processes within the Earth's crust.

The structure consists of concentric rings of sedimentary rock, which have been exposed and eroded over millions of years. The layers of rock, each representing different geological periods, create the bullseye pattern that is visible today. The central part of the structure is composed of older rocks, while the outer rings are made up of younger sedimentary layers. This pattern provides valuable insights into the geological history of the region and the processes that shaped it.

The formation of the Richat Structure is believed to have begun around 100 million years ago during the Cretaceous period, when volcanic activity caused the Earth's crust to bulge upward, creating a dome. Over time, erosion wore away the upper layers of the dome, exposing the underlying rock and creating the concentric rings. The exact mechanisms behind the formation of the dome are

161

still a subject of scientific research, with theories suggesting that it may have been caused by magma intrusion or tectonic forces.

The Eye of the Sahara is not only a geological wonder but also a site of archaeological interest. The area surrounding the structure contains evidence of ancient human habitation, including stone tools and other artifacts. These findings suggest that early humans lived in the region when it was more hospitable, before the Sahara Desert became the arid landscape it is today. The Richat Structure's unique features may have made it a significant landmark for these ancient inhabitants.

In addition to its scientific and archaeological importance, the Eye of the Sahara has captured the imagination of people around the world. Its unusual appearance has led to various myths and speculations, including theories that it may be the remnants of the lost city of Atlantis or other ancient civilizations. While there is no evidence to support these claims, the structure's enigmatic beauty continues to inspire curiosity and wonder.

The Eye of the Sahara is also a popular destination for adventurous travelers and photographers. Its remote location and striking visual appeal make it a unique and memorable sight for those who venture to the Sahara Desert. The structure's concentric rings and vibrant colors create a surreal landscape that is unlike anything else on Earth.

Conservation efforts are important to preserve the Eye of the Sahara and its surrounding environment. The Sahara Desert is a fragile ecosystem, and human activities such as mining, tourism, and climate change can have significant impacts on the region. Protecting the Richat Structure and its geological and archaeological features is essential for future generations to appreciate and study.

In conclusion, the Eye of the Sahara, or the Richat Structure, is a remarkable geological formation that stands out as a natural wonder in the vast expanse of the Sahara Desert. Its concentric rings, created by millions of years of geological processes, offer valuable insights into the Earth's history and the forces that shape our planet. As a site of scientific, archaeological, and cultural significance, the Eye of the Sahara continues to fascinate and inspire people around the world.

10. The Aurora Australis

The Aurora Australis, also known as the Southern Lights, is the southern counterpart to the Aurora Borealis. This dazzling natural light display occurs near the Earth's South Pole, primarily visible from high-latitude regions such as Antarctica, New Zealand, and the southern parts of Australia. The Aurora Australis shares the same fundamental mechanisms as the Aurora Borealis, resulting in spectacular light shows that illuminate the night sky with vibrant colors and dynamic patterns.

The phenomenon of the Aurora Australis is caused by the interaction between the solar wind and the Earth's magnetosphere. When charged particles from the sun, primarily electrons and protons, collide with the Earth's magnetic field, they are directed toward the polar regions. As these particles enter the Earth's upper atmosphere, they collide with oxygen and nitrogen molecules, releasing energy in the form of light. The colors observed in the aurora are determined by the type of gas and the altitude of the collisions.

Oxygen atoms at higher altitudes (above 150 miles) produce red auroras, while those at lower altitudes (up to 150 miles) emit green light, the most common color seen in the aurora. Nitrogen molecules can produce blue or purplish-red auroras, depending on the energy of the collisions. The result is a breathtaking display of colors that dance across the sky, creating an ever-changing and mesmerizing spectacle.

The best time to observe the Aurora Australis is during the austral winter months, from March to September, when the nights are longest and the skies are darkest. Areas with minimal light pollution, such as remote parts of Tasmania, New Zealand's South Island, and, of course, Antarctica, offer the best viewing conditions. During periods of high solar activity, the aurora can sometimes be seen from lower latitudes, providing a rare and enchanting experience for observers.

The Aurora Australis has been a subject of fascination and reverence for many cultures throughout history. Indigenous peoples in the Southern Hemisphere, such as the Maori of New Zealand and the Aboriginal Australians, have rich traditions and legends associated with the Southern Lights. These stories often depict the aurora as a manifestation of the spirits of ancestors or as a celestial phenomenon imbued with spiritual significance.

Scientific exploration of the Aurora Australis has provided valuable insights into space weather and the Earth's magnetosphere. Early observations and studies of the aurora were conducted by explorers and scientists during expeditions to the polar regions. With the advent of modern technology, satellites and ground-based observatories equipped with advanced instruments have allowed scientists to monitor and study the aurora in greater detail. These observations help researchers understand the dynamics of the Earth's magnetic field, the behavior of the solar wind, and the potential impacts of space weather on our planet.

The study of auroras has practical applications as well. Understanding the mechanisms behind auroral activity can improve our ability to predict space weather events, such as geomagnetic storms, which can affect satellite communications, navigation systems, and power grids. By studying the auroras, scientists can develop better models and forecasting tools to mitigate the potential impacts of space weather on technology and infrastructure.

The Aurora Australis is not only a subject of scientific interest but also a major attraction for tourists and photographers. Each year, thousands of people travel to southern latitudes to witness the stunning light displays. Tourism industries in regions like Tasmania and New Zealand have developed guided tours, accommodations, and educational programs to enhance the aurora-viewing experience. These efforts help raise awareness about the natural beauty and scientific significance of the auroras while supporting local economies.

In conclusion, the Aurora Australis is a spectacular natural phenomenon that illuminates the night skies of the Southern Hemisphere with vibrant colors and dynamic patterns. This breathtaking light show, caused by the interaction between the solar wind and the Earth's magnetosphere, shares the same fundamental mechanisms as the Aurora Borealis. The Aurora Australis captivates and inspires observers, offering a glimpse into the dynamic and interconnected nature of our planet and its space environment. As we continue to study and appreciate the auroras, these magnificent light displays remain a powerful reminder of the beauty and complexity of the natural world.

Chapter 8: Unusual Weather Phenomena

1. Ball Lightning

Ball lightning is one of the most mysterious and rare weather phenomena observed throughout history. Described as glowing, spherical objects that can appear during thunderstorms, ball lightning has intrigued and baffled scientists and laypeople alike. Despite numerous reports and anecdotal evidence, the exact nature and cause of ball lightning remain subjects of ongoing scientific investigation.

The phenomenon of ball lightning typically manifests as a luminous, floating orb, ranging in size from a few centimeters to several meters in diameter. These glowing balls can be white, yellow, orange, red, or blue, and they often move erratically, sometimes hovering or bouncing before disappearing, either quietly or with an explosive sound. Reports of ball lightning have come from various parts of the world, with observations dating back centuries.

One of the earliest recorded descriptions of ball lightning comes from the English physician and philosopher John Stow, who in 1596 described a "great ball of fire" that entered a church and caused considerable damage. Since then, numerous accounts have been documented, including those from prominent scientists such as Nikola Tesla and Winston Churchill. However, the sporadic and unpredictable nature of ball lightning has made it difficult to study systematically.

Several theories have been proposed to explain ball lightning, but none have been universally accepted. Some of the leading hypotheses include:

Vaporized Silicon Hypothesis: One theory suggests that ball lightning is caused by the vaporization of silicon in soil during a lightning strike. The vaporized silicon reacts with oxygen in the air, forming a glowing, silicon-based plasma ball. This hypothesis is supported by laboratory experiments that have successfully created luminous, ball-like objects using silicon.

Plasma Torus Hypothesis: Another theory proposes that ball lightning is a form of plasma, a state of matter consisting of ionized gas. This plasma could be trapped in a toroidal, or doughnut-shaped, magnetic field, creating a stable, glowing sphere. However, the specific conditions required to produce and sustain such a plasma torus are not well understood.

Microwave Cavity Hypothesis: Some researchers believe that ball lightning may be caused by microwaves produced during a lightning strike. These microwaves could become trapped in a cavity, such as a sphere of ionized air, creating a stable, glowing ball of energy. This theory is supported by observations of ball lightning appearing near metal objects, which could act as microwave reflectors.

Chemical Reactions Hypothesis: Another possibility is that ball lightning results from complex chemical reactions between gases in the atmosphere. These reactions could produce a glowing, stable ball of gas that persists for several seconds to minutes. However, the exact chemical processes involved are not well understood.

Despite the lack of a definitive explanation, advances in technology and research methods have provided new opportunities to study ball lightning. High-speed cameras, spectrometers, and other scientific instruments have been used to capture and analyze rare instances of ball lightning, providing valuable data on its properties and behavior.

Ball lightning has also been recreated in laboratory settings, offering insights into the potential mechanisms behind the phenomenon. These experiments have demonstrated that it is possible to produce glowing, spherical objects under controlled conditions, lending credence to some of the proposed theories.

In addition to scientific interest, ball lightning has captured the public imagination and inspired numerous cultural references. It has been featured in literature, folklore, and popular media, often portrayed as a mysterious and awe-inspiring natural phenomenon. The enduring fascination with ball lightning reflects humanity's broader curiosity about the natural world and the desire to understand the unknown.

As research continues, the study of ball lightning may not only reveal the secrets of this elusive phenomenon but also contribute to our understanding of plasma physics, atmospheric science, and the behavior of electrical discharges. Unraveling the mystery of ball lightning could have practical applications in fields such as energy production, weather prediction, and lightning protection.

In conclusion, ball lightning remains one of the most intriguing and enigmatic weather phenomena. Despite centuries of observations and numerous theories, its exact nature and origin continue to elude definitive explanation. Ongoing scientific research and technological advancements hold the promise of unlocking the secrets of ball lightning, deepening our knowledge of this captivating natural occurrence.

2.   Fire Whirls

Fire whirls, also known as fire tornadoes or fire devils, are a rare and dramatic weather phenomenon that occurs during intense wildfires. These spinning vortices of flame can reach heights of several hundred feet and exhibit destructive power akin to that of traditional tornadoes. The formation and behavior of fire whirls are driven by complex interactions between fire, wind, and atmospheric conditions, making them a subject of significant scientific interest and a striking example of nature's fury.

A fire whirl typically forms when a wildfire generates a strong updraft of hot air. As the hot air rises, cooler air rushes in to replace it, creating a rotating column of air. If the conditions are right, this rotating column can become tightly focused, forming a vortex that pulls in flames, ash, and debris. The intense heat within the vortex can cause the flames to burn more fiercely, producing a spinning column of fire that can last for several minutes.

The key factors contributing to the formation of a fire whirl include:

Intense Heat: A large, intense wildfire generates significant heat, creating strong updrafts that are necessary for the formation of a fire whirl. The heat also increases the buoyancy of the air, enhancing the upward movement.

Wind Shear: Variations in wind speed and direction at different altitudes, known as wind shear, can induce rotation in the rising column of hot air. This rotation is essential for the development of a vortex.

Topography: The shape and features of the landscape can influence the formation of fire whirls. Hills, valleys, and other landforms can channel and accelerate wind currents, contributing to the spinning motion.

Atmospheric Instability: Conditions of atmospheric instability, where the temperature decreases rapidly with altitude, can enhance the vertical movement of air, promoting the development of fire whirls.

Fire whirls can vary in size and intensity. Smaller fire whirls are relatively common in large wildfires and may last only a few seconds, causing localized damage. Larger, more intense fire whirls, sometimes called fire tornadoes, can reach several hundred feet in height and exhibit wind speeds exceeding 100 miles per hour. These powerful fire whirls can uproot trees, toss vehicles, and spread flames over long distances, exacerbating the wildfire and posing significant threats to firefighters and nearby communities.

One of the most notable examples of a fire whirl occurred during the 1923 Great Kanto Earthquake in Japan. The earthquake triggered a massive firestorm in Tokyo, and the intense heat and chaotic wind conditions produced a fire whirl that swept through the city, causing widespread devastation and claiming thousands of lives. This tragic event highlighted the destructive potential of fire whirls and underscored the need for understanding and mitigating their impact.

Modern firefighting strategies take into account the possibility of fire whirls when battling large wildfires. Firefighters are trained to recognize the signs of developing fire whirls and to adjust their tactics accordingly to minimize the risk. Understanding the conditions that lead to the formation of fire whirls can help in predicting their occurrence and improving the safety and effectiveness of firefighting efforts.

Research into fire whirls continues to advance, with scientists using both field observations and laboratory experiments to study their formation and behavior. High-speed cameras, thermal imaging, and computer simulations provide valuable data on the dynamics of fire whirls, helping researchers develop more accurate models and predictive tools. These studies not only contribute to wildfire management but also enhance our broader understanding of fluid dynamics and vortex behavior.

In addition to their scientific and practical significance, fire whirls capture the imagination due to their dramatic and awe-inspiring appearance. The sight of a towering column of flame spinning rapidly against the backdrop of a wildfire is a powerful reminder of nature's capacity for both beauty and destruction.

In conclusion, fire whirls are a rare and intense weather phenomenon that occurs during large wildfires. Their formation is driven by complex interactions between heat, wind, topography, and atmospheric conditions. Understanding fire whirls is crucial for improving wildfire management and ensuring the safety of firefighters and communities. As research continues, these fiery vortices remain a striking example of the dynamic and unpredictable forces at play in our natural world.

3. Blood Rain

Blood rain is a rare and striking weather phenomenon in which rain appears red or reddish-brown, giving it the appearance of falling blood. This unusual occurrence has been documented throughout history and has often been attributed to supernatural or ominous causes. However, modern science has provided a clear explanation for the phenomenon, revealing that blood rain is the result of airborne particles, such as dust or pollen, mixing with rainwater and imparting a reddish hue.

The primary cause of blood rain is the presence of high concentrations of dust or other fine particles in the atmosphere. These particles can originate from various sources, including desert dust storms, volcanic ash, and industrial pollution. When these particles are lifted into the atmosphere by strong winds, they can be carried over long distances. If rain occurs while these particles are present in the atmosphere, they can become suspended in the raindrops, giving the rain a red or brown color.

One of the most well-known sources of blood rain is Saharan dust. Each year, large quantities of dust are lifted from the Sahara Desert in North Africa and transported across the Mediterranean Sea and even as far as Europe and the Americas. When this dust coincides with rain, it can produce the characteristic reddish-brown rain that has been described in numerous historical accounts.

In addition to desert dust, blood rain can also be caused by the presence of microorganisms, such as algae or spores, in the atmosphere. For example, a species of microalgae called Trentepohlia can produce a red pigment. When these algae are

171

airborne and mix with rain, they can create the appearance of blood rain. Similarly, spores from certain fungi can also give rain a reddish tint.

The phenomenon of blood rain has been recorded throughout history and across different cultures. In ancient and medieval times, blood rain was often interpreted as a bad omen or a sign of impending disaster. Historical records from Europe, India, and other regions describe instances of blood rain that were accompanied by fear and superstition. Some believed that the rain was literally composed of blood and saw it as a divine warning or a harbinger of war, plague, or other calamities.

Modern scientific investigations have dispelled these superstitions, providing a clear understanding of the natural causes behind blood rain. Studies using spectroscopy and chemical analysis have shown that the red coloration in blood rain is due to the presence of iron oxides and other mineral particles. These particles are commonly found in desert dust and can impart a red or brown color when suspended in water.

Blood rain events continue to capture public interest and curiosity. In recent years, instances of blood rain have been reported in places such as India, Sri Lanka, and parts of Europe. These events are often associated with Saharan dust storms or other sources of atmospheric particles. While blood rain may appear alarming, it poses no significant threat to human health. The particles responsible for the coloration are typically harmless, though they can contribute to air quality issues when present in high concentrations.

In conclusion, blood rain is a rare and fascinating weather phenomenon caused by the mixing of airborne particles with rainwater. The particles, often originating from desert dust or microorganisms, give the rain a distinctive red or brown color. While blood rain has historically been associated with superstition and fear, modern science has provided a clear and rational explanation for this natural occurrence. As we continue to study and understand atmospheric processes, phenomena like blood rain serve as a reminder of the intricate and dynamic interactions between the Earth's atmosphere and the environment.

4. The Green Flash

The green flash is a rare and captivating optical phenomenon that occurs shortly after sunset or just before sunrise. Observers at the right place and time may witness a brief, green-colored spot of light appear on the upper edge of the sun's disk. This fleeting event, lasting only a few seconds, is a result of atmospheric refraction and dispersion of sunlight, creating a memorable and sought-after sight for sky watchers and photographers.

The science behind the green flash lies in the behavior of light as it passes through the Earth's atmosphere. Light from the sun is composed of different colors, each with a slightly different wavelength. When sunlight enters the Earth's atmosphere, it is refracted, or bent, by the varying density of the air. The amount of bending depends on the wavelength of the light; shorter wavelengths (blue and green light) are refracted more than longer wavelengths (red and orange light).

As the sun sets or rises, its light must pass through a greater thickness of the Earth's atmosphere compared to when it is higher in the sky. This increased distance causes more significant refraction and dispersion of the sunlight. During this process, the different colors of light are separated, with the green and blue light appearing slightly higher in the sky than the red and orange light.

The green flash occurs during the final moments of sunset or the initial moments of sunrise, when the sun's disk is just below the horizon. The atmosphere acts like a prism, separating the colors and allowing the upper rim of the sun to briefly display a green hue.

173

The phenomenon is fleeting because the angle of refraction changes rapidly as the sun continues to set or rise, causing the green light to disappear almost as quickly as it appears.

Several factors influence the likelihood of observing a green flash. Clear skies with minimal atmospheric turbulence and a flat, unobstructed horizon, such as the ocean or a wide plain, provide the best conditions. The presence of atmospheric particles, such as dust or pollution, can also affect the intensity and visibility of the green flash. Observers at higher altitudes, such as mountain tops, may have a better chance of seeing the phenomenon due to the thinner atmosphere.

There are different types of green flashes, varying in appearance and duration. The most common type is the "inferior mirage" green flash, which occurs when the air near the surface is warmer than the air above it, creating a mirage effect that enhances the refraction of light. Another type is the "mock mirage" green flash, which occurs under temperature inversion conditions, where a layer of cooler air is trapped beneath a layer of warmer air. This type can produce a more pronounced and longer-lasting green flash.

While the green flash is a well-documented phenomenon, it has also inspired numerous myths and legends. Sailors and coastal communities have long regarded the green flash as a good omen or a sign of fair weather. In literature, the green flash has been romanticized as a magical or mystical event. Jules Verne's novel "Le Rayon Vert" ("The Green Ray") popularized the phenomenon in the 19th century, portraying it as a moment of perfect clarity and vision.

Despite its rarity, the green flash has become a popular target for photographers and sky watchers. Capturing the green flash on camera requires precise timing and favorable conditions, making it a challenging yet rewarding endeavor. High-quality optics, such as binoculars or telescopes, can enhance the viewing experience and increase the chances of witnessing this elusive event.

In conclusion, the green flash is a rare and beautiful optical phenomenon resulting from the atmospheric refraction and dispersion of sunlight during sunrise or sunset. While it lasts only a few seconds, the green flash offers a glimpse into the complex interactions between light and the Earth's atmosphere. For those fortunate enough to witness it, the green flash is a memorable and awe-inspiring reminder of the natural wonders that occur in our everyday world.

5. Ice Circles

Ice circles, also known as ice discs or ice pans, are rare and intriguing natural phenomena that occur in slow-moving rivers and streams during cold weather conditions. These perfectly round or nearly round discs of ice can range in size from a few centimeters to several meters in diameter, and they slowly rotate as they float on the water's surface. Ice circles captivate observers with their unusual appearance and the seemingly delicate balance that allows them to form and persist.

The formation of ice circles is influenced by a combination of environmental factors, including water temperature, flow velocity, and rotational currents. The primary mechanism behind the creation of ice circles involves the following steps:

Initial Ice Formation: Ice circles typically begin to form when the air temperature drops and a thin layer of ice starts to develop on the surface of slow-moving water. This initial ice layer may form naturally or be a result of ice chunks breaking off from the riverbanks and floating downstream.

Rotational Currents: The key to the formation of an ice circle is the presence of a rotational current, which can be caused by an obstruction in the river, such as a rock or a bend in the waterway.

175

The rotational current causes the floating ice to begin spinning slowly. As the ice rotates, it grinds against other ice pieces and the riverbanks, gradually becoming more circular.

Erosion and Growth: The rotation of the ice disc causes it to erode along the edges, smoothing out irregularities and creating a more uniform shape. The circular motion also helps to gather additional ice and snow from the water's surface, allowing the disc to grow in size over time.

Stable Equilibrium: Once the ice circle reaches a stable equilibrium, it can persist for several hours or even days, depending on the environmental conditions. The slow rotation helps to maintain the disc's shape and prevents it from becoming lodged against the riverbanks.

While ice circles are most commonly found in rivers and streams, they can also form in lakes and ponds under the right conditions. The phenomenon has been observed in various parts of the world, including North America, Scandinavia, and Russia. Ice circles are more likely to form in regions with cold climates and slow-moving or stagnant water bodies.

The first documented observation of an ice circle dates back to the 19th century, and since then, the phenomenon has been reported sporadically in scientific literature and news media. One of the most famous instances occurred in 2008 in the Sheyenne River in North Dakota, where a large ice circle measuring about 15 meters (50 feet) in diameter attracted widespread attention and curiosity.

Despite their rare occurrence, ice circles have sparked the interest of scientists, who study them to understand the complex interactions between water flow, temperature, and ice formation. The phenomenon provides insights into fluid dynamics, thermodynamics, and the behavior of natural systems under varying environmental conditions.

In addition to their scientific significance, ice circles have captured the public's imagination and inspired a sense of wonder and curiosity. Their symmetrical beauty and seemingly perfect shape make them popular subjects for photography and nature observation. Social media and online platforms have played a significant role in sharing images and videos of ice circles, further raising awareness and interest in this fascinating natural phenomenon.

Ice circles also hold cultural significance in some regions, where they are viewed as symbols of harmony and balance in nature. Their ephemeral nature and the conditions required for their formation serve as reminders of the delicate interplay between natural forces and the ever-changing environment.

In conclusion, ice circles are rare and mesmerizing natural phenomena that form in slow-moving rivers and streams during cold weather conditions. The combination of initial ice formation, rotational currents, erosion, and growth results in perfectly round or nearly round discs of ice that float and rotate on the water's surface. Ice circles offer valuable insights into the dynamics of natural systems and continue to captivate and inspire those fortunate enough to witness them.

6. Mammatus Clouds

Mammatus clouds, often referred to as mammatocumulus, are distinctive cloud formations characterized by their pouch-like, bulging underbellies. These clouds are typically associated with severe thunderstorms and are often seen trailing behind cumulonimbus clouds. Mammatus clouds are both visually striking and scientifically intriguing, serving as a dramatic indicator of turbulent atmospheric conditions.

The term "mammatus" is derived from the Latin word "mamma," meaning "udder" or "breast," which aptly describes the cloud's appearance. Each pouch or lobe of a mammatus cloud can vary in size, ranging from small, subtle protrusions to large, pronounced bulges that extend for miles across the sky. The unique structure of these clouds can create a surreal and almost otherworldly sky, captivating the attention of observers.

The formation of mammatus clouds involves a combination of atmospheric instability, moisture, and sinking air currents. Unlike most clouds that form due to rising air, mammatus clouds develop through a descending process. Here is a simplified explanation of how mammatus clouds form:

Instability and Moisture: Mammatus clouds are typically found in environments with strong atmospheric instability and high moisture content. These conditions are often present in the anvil or cirrus shield of a mature thunderstorm.

Cooling and Sinking Air: Within the anvil of the thunderstorm, pockets of cooler, denser air can form. As this air descends, it sinks into the warmer, less dense air below, creating the characteristic pouch-like structures. This process is driven by negative buoyancy, where the cooler air is heavier than the surrounding air.

Condensation and Evaporation: The sinking air can lead to the condensation of water vapor, forming the visible cloud pouches. The interplay between condensation and evaporation at the boundaries of these pouches can enhance their distinct appearance, creating sharp edges and well-defined lobes.

Mammatus clouds are most commonly associated with severe thunderstorms and supercell storms. They can also occur in the vicinity of other intense weather systems, such as tropical cyclones and volcanic eruptions. While the presence of mammatus clouds often indicates severe weather, they themselves are not harmful. However, their appearance can serve as a visual warning of the potential for hazardous conditions, such as heavy rainfall, hail, strong winds, and lightning.

In addition to their meteorological significance, mammatus clouds have captured the interest of photographers and sky watchers. Their dramatic and unusual appearance makes them a popular subject for photography, especially during sunset or sunrise when the clouds can be illuminated with vibrant colors. Images of mammatus clouds often evoke a sense of awe and wonder, highlighting the dynamic and ever-changing nature of the Earth's atmosphere.

Scientific research on mammatus clouds continues to advance our understanding of atmospheric processes and cloud dynamics. By studying these clouds, meteorologists can gain insights into the behavior of severe storms and improve weather forecasting models. High-resolution satellite imagery and advanced radar systems have enhanced the ability to observe and analyze mammatus cloud formations, contributing to the broader field of cloud physics.

Mammatus clouds also hold cultural significance in various societies. They have been interpreted as omens or signs in folklore and mythology, often seen as portents of storms or other significant events. Their striking appearance has inspired artistic representations and has been used symbolically in literature and media to evoke a sense of impending drama or transformation.

In conclusion, mammatus clouds are distinctive and visually captivating cloud formations associated with severe thunderstorms and atmospheric turbulence. Their unique pouch-like structures result from sinking air currents and atmospheric instability, creating a dramatic and often awe-inspiring sky. While they serve as indicators of potential severe weather, mammatus clouds themselves are a testament to the complexity and beauty of the Earth's atmosphere, continuing to inspire curiosity and admiration among scientists and sky watchers alike.

7. Lenticular Clouds

Lenticular clouds, also known as lenticularis, are unique cloud formations that resemble smooth, lens-shaped discs or saucers. These stationary clouds often form over mountain ranges or other geographic features that disrupt airflow, creating the perfect conditions for their development. Lenticular clouds are not only visually striking but also provide valuable insights into atmospheric dynamics and wave patterns.

The formation of lenticular clouds is closely linked to the presence of stable, moist air and topographic features that induce wave-like motion in the atmosphere. Here is a simplified explanation of how lenticular clouds form:

Orographic Lifting: As moist, stable air flows over a mountain or hill, it is forced to ascend, cooling as it rises. This process, known as orographic lifting, creates a wave-like motion in the atmosphere. The air cools and condenses as it reaches the crest of the wave, forming clouds.

Wave Formation: When the air descends on the leeward side of the mountain, it warms and evaporates, creating clear areas. This ascending and descending motion can set up a series of standing waves downwind from the mountain, with each wave crest potentially forming a lenticular cloud if the air remains moist enough.

Cloud Stability: Lenticular clouds are typically stable and remain fixed in position relative to the geographic feature causing the wave. They can persist for hours or even days, as long as the flow of air and moisture remains consistent.

Lenticular clouds are most commonly observed in mountainous regions but can also form in other areas where the right conditions for atmospheric waves exist. They are often seen in places like the Rocky Mountains in the United States, the Andes in South

America, the Alps in Europe, and the Southern Alps in New Zealand.

One of the most fascinating aspects of lenticular clouds is their smooth, saucer-like appearance. These clouds can stack on top of one another, creating a multi-layered effect that resembles a stack of pancakes or a series of lenses. Their unique shape and stationary nature make them easily distinguishable from other cloud types.

Pilots and aviation enthusiasts pay close attention to lenticular clouds because they are indicators of strong atmospheric turbulence. The wave patterns associated with these clouds can create severe updrafts and downdrafts, posing challenges for aircraft navigation. Glider pilots, however, often seek out lenticular clouds, as the strong updrafts can provide excellent lift for long-distance soaring.

Lenticular clouds have also inspired numerous cultural references and myths. Due to their unusual and often otherworldly appearance, they have been linked to UFO sightings and extraterrestrial activity. Their smooth, disc-like shape has led some people to mistake them for flying saucers, fueling speculation and intrigue.

In addition to their visual appeal and cultural significance, lenticular clouds offer valuable insights into atmospheric wave patterns and fluid dynamics. Scientists study these clouds to understand the interactions between airflow and topography, which can have implications for weather forecasting, climate modeling, and aviation safety.

Photographers and sky watchers are particularly drawn to lenticular clouds due to their dramatic and picturesque appearance. Capturing images of these clouds against the backdrop of a mountain range or at sunset, when they can be illuminated with vibrant colors, often results in stunning and memorable photographs. Social media and online platforms have helped to share and popularize these images, raising awareness of the beauty and significance of lenticular clouds.

In conclusion, lenticular clouds are unique and visually striking cloud formations that form over mountains and other geographic features. Their smooth, lens-like shape and stationary nature make them easily recognizable and a subject of fascination for scientists, pilots, and sky watchers alike. By studying lenticular clouds, researchers gain valuable insights into atmospheric wave patterns and fluid dynamics, while their dramatic appearance continues to inspire cultural references and photographic pursuits. Lenticular clouds stand as a testament to the intricate and dynamic interactions within our planet's atmosphere, showcasing the beauty and complexity of natural phenomena.

8. Light Pillars

Light pillars are a captivating atmospheric phenomenon that creates vertical columns of light extending above or below a light source. This optical event occurs under specific conditions and can be observed at night or during twilight. Light pillars are caused by the reflection of light from numerous tiny ice crystals suspended in the atmosphere. These shimmering columns of light are often seen in colder regions and can be produced by both natural and artificial light sources.

The formation of light pillars involves the following key factors:

Ice Crystals: Light pillars are formed when light reflects off the flat surfaces of hexagonal plate-like ice crystals in the atmosphere. These ice crystals are typically found in cirrostratus or altostratus clouds but can also be present near the ground in very cold conditions, forming in ice fog or diamond dust.

Reflection: When light from a source such as the sun (during sunrise or sunset), the moon, or artificial lights (such as streetlights) interacts with the ice crystals, the light is reflected vertically. The flat, hexagonal surfaces of the ice crystals act like tiny mirrors, reflecting the light upward or downward to create a vertical column.

Alignment: For light pillars to be visible, the ice crystals must be relatively well-aligned with their flat faces parallel to the ground. This alignment allows the light to reflect in a way that creates the appearance of a pillar extending above or below the light source.

Light pillars can occur in a variety of settings and can be categorized based on their light source:

Solar Pillars: These light pillars are created by the reflection of sunlight from ice crystals. They are typically seen during sunrise or sunset when the sun is low on the horizon. Solar pillars can appear as a glowing column of light extending upward or downward from the sun.

Lunar Pillars: Similar to solar pillars, lunar pillars are formed by the reflection of moonlight from ice crystals. These are often seen on clear, cold nights when the moon is near the horizon and can produce a serene and ethereal display.

Artificial Light Pillars: In urban areas, light pillars can be created by artificial light sources such as streetlights, stadium lights, or other bright lights. These pillars are often seen in colder regions where ice crystals are present near the ground, reflecting the artificial lights to create vertical columns.

Light pillars are most commonly observed in polar and temperate regions during winter months when temperatures are low enough to support the formation of ice crystals in the atmosphere. Locations such as Alaska, Canada, Scandinavia, and Russia frequently report sightings of light pillars, particularly during cold, clear nights.

The visual spectacle of light pillars has captured the interest of photographers and sky watchers. The vibrant and surreal appearance of these columns of light makes them a popular subject

183

for night photography. Capturing light pillars requires clear skies, cold temperatures, and a bit of patience, as the phenomenon can be fleeting and depends on specific atmospheric conditions.

In addition to their aesthetic appeal, light pillars offer insights into atmospheric optics and the behavior of light. Studying light pillars helps scientists understand the properties and alignment of ice crystals in the atmosphere, as well as the conditions that lead to their formation. This knowledge can contribute to broader research in meteorology and climatology.

Light pillars have also found their way into cultural references and folklore. Their appearance has been interpreted in various ways, from omens and supernatural signs to symbols of beauty and mystery. The phenomenon's striking and otherworldly appearance continues to inspire wonder and curiosity.

In conclusion, light pillars are a remarkable and visually stunning atmospheric phenomenon caused by the reflection of light from ice crystals. Whether created by sunlight, moonlight, or artificial lights, these vertical columns of light offer a captivating display that enchants observers and provides valuable insights into atmospheric processes. Light pillars serve as a reminder of the intricate and dynamic interactions within our planet's atmosphere, showcasing the beauty and complexity of natural phenomena.

9. Moonbows

Moonbows, also known as lunar rainbows, are a rare and enchanting optical phenomenon that occurs when moonlight, rather than sunlight, is refracted and reflected by water droplets in the atmosphere. These nocturnal rainbows are typically fainter than their daytime counterparts due to the lower intensity of moonlight, but they possess a subtle and ethereal beauty that makes them a sought-after sight for sky watchers and photographers.

The formation of a moonbow involves the same basic principles as a solar rainbow, but with moonlight as the light source. The process can be summarized as follows:

Moonlight: Moonbows are produced by the light from a nearly full moon. The moon must be bright enough to produce the necessary light intensity for a rainbow to form. Typically, moonbows are seen within a few days before, during, or after a full moon.

Water Droplets: As with solar rainbows, moonbows require the presence of water droplets in the atmosphere, such as those found in rain, mist, or the spray from waterfalls. The droplets act as tiny prisms, refracting, reflecting, and dispersing the moonlight to create the rainbow.

Angle of Reflection: The angle at which the moonlight enters and exits the water droplets determines the formation of the rainbow. For a moonbow to be visible, the observer must be positioned with the moon behind them, and the water droplets must be in front of them, with the light reflecting back at an angle of 42 degrees.

The faintness of moonbows compared to solar rainbows is due to the fact that moonlight is much less intense than sunlight. As a result, moonbows often appear white or colorless to the naked eye, as the colors are not bright enough to be easily distinguished. However, with long-exposure photography, the full spectrum of colors can be captured, revealing the moonbow's hidden beauty.

Moonbows are most commonly observed in areas with frequent rainfall and clear night skies. Some of the best-known locations for observing moonbows include:

Cumberland Falls, Kentucky, USA: Known as the "Niagara of the South," Cumberland Falls is famous for its regular occurrence of moonbows, which can be seen on clear nights with a full moon. The falls' powerful spray provides the ideal conditions for the formation of lunar rainbows.

Victoria Falls, Zambia/Zimbabwe: One of the largest and most famous waterfalls in the world, Victoria Falls frequently produces moonbows due to its immense volume of water and mist. Moonbows at Victoria Falls can often be seen during the full moon and are a major attraction for tourists.

Waimea, Hawaii, USA: The waterfalls and frequent rain showers in Waimea make it another prime location for observing moonbows. The combination of tropical climate and clear night skies provides ideal conditions for this phenomenon.

Yosemite National Park, California, USA: Yosemite's Bridalveil Fall and other waterfalls in the park are known for producing moonbows, especially during the spring and early summer when water flow is at its peak.

While moonbows are a natural wonder, they are also subject to the whims of weather and atmospheric conditions. Factors such as cloud cover, air pollution, and light pollution can affect the visibility and intensity of moonbows. Clear, dark skies with minimal artificial light are essential for the best viewing experience.

Photographers seeking to capture moonbows often use long-exposure techniques to bring out the colors and details of the rainbow. Tripods, wide-angle lenses, and proper exposure settings are crucial for achieving the desired effect. Capturing a moonbow on camera requires patience and a bit of luck, but the resulting images can be breathtakingly beautiful.

In addition to their aesthetic appeal, moonbows provide insights into the physics of light and the interactions between light and water droplets. The study of moonbows and other atmospheric optics helps scientists understand the behavior of light in different conditions and contributes to the broader field of meteorology.

In conclusion, moonbows are a rare and captivating phenomenon that occurs when moonlight interacts with water droplets to create a faint, ethereal rainbow. While they are less common and more challenging to observe than solar rainbows, moonbows offer a unique and enchanting experience for those fortunate enough to witness them. Whether seen in person or captured through the lens of a camera, moonbows serve as a reminder of the subtle and delicate beauty that can be found in the natural world.

10. Frost Flowers

Frost flowers are delicate, ice crystal formations that resemble blooming flowers, created under specific and rare atmospheric conditions. These stunning natural sculptures typically form on the surface of sea ice, lake ice, or even on the stems of plants during cold, calm weather. Frost flowers are not only visually captivating but also provide valuable insights into the processes of ice formation and atmospheric interactions.

The formation of frost flowers involves a combination of moisture, cold temperatures, and calm conditions. Here's how they typically form:

Cold, Calm Conditions: Frost flowers form in extremely cold temperatures, usually below -22°F (-30°C). Calm and stable weather conditions are crucial, as wind can disrupt the delicate formation process.

Moisture Source: Frost flowers require a source of moisture, which can come from sea ice, lake ice, or even plant stems. On sea ice, the moisture is provided by the supersaturated air just above the ice surface, which is rich in water vapor. This vapor can originate from small cracks or leads in the ice, where relatively warmer seawater comes into contact with the cold air.

Ice Crystal Formation: As the moisture-laden air comes into contact with the cold surface, the water vapor condenses and freezes, forming intricate ice crystals. These crystals can grow rapidly, creating the intricate and delicate structures characteristic of frost flowers. On plant stems, the moisture is drawn up from

187

the roots and extruded through the stem, where it freezes upon contact with the cold air.

Continuous Growth: Once the initial ice crystals form, they can continue to grow as long as the conditions remain stable. The result is a cluster of ice crystals that resemble delicate petals or flowers.

Frost flowers are most commonly observed in polar and subpolar regions, where the necessary conditions for their formation are more likely to occur. They have been documented in places such as the Arctic, Antarctica, and high-altitude mountain ranges. However, frost flowers can also form in temperate regions under the right conditions, such as during cold snaps or on frozen lakes and rivers.

These delicate ice formations are not only beautiful but also scientifically significant. Frost flowers can influence the local atmosphere by releasing salts and other chemicals trapped within the ice into the air. This process can affect the composition of the boundary layer of the atmosphere and contribute to the formation of aerosols. Additionally, the presence of frost flowers can provide a habitat for certain types of microorganisms, which can survive within the ice structures.

Frost flowers also play a role in the study of climate change and polar environments. Researchers study these formations to understand the interactions between sea ice, atmospheric conditions, and the broader climate system. The growth and sublimation of frost flowers can impact the albedo, or reflectivity, of the ice surface, influencing how much solar energy is absorbed or reflected.

Photographers and nature enthusiasts are often captivated by the ephemeral beauty of frost flowers. Capturing these delicate structures on camera requires careful timing and an understanding of the environmental conditions that support their formation. The

188

intricate details and otherworldly appearance of frost flowers make them a popular subject for winter photography.

In addition to their scientific and aesthetic appeal, frost flowers have found their way into cultural references and folklore. In some cultures, they are seen as symbols of purity and the fleeting beauty of nature. Their brief existence, often lasting only a few hours before sublimating back into the atmosphere, serves as a reminder of the transient nature of many natural phenomena.

In conclusion, frost flowers are rare and exquisite ice formations that occur under specific atmospheric conditions. Their delicate, flower-like appearance and ephemeral nature make them a captivating subject for both scientific study and artistic appreciation. By understanding the processes behind their formation, researchers gain valuable insights into ice dynamics, atmospheric interactions, and the impacts of climate change. Frost flowers stand as a testament to the intricate and delicate beauty that can be found in the natural world, even in the harshest and coldest environments.

Chapter 9: Astonishing Geological Formations

1. Giant's Causeway

Giant's Causeway, located on the northeast coast of Northern Ireland, is one of the most striking and iconic geological formations in the world. This UNESCO World Heritage Site is renowned for its approximately 40,000 interlocking basalt columns, which were formed by volcanic activity around 60 million years ago. The columns, predominantly hexagonal in shape, create a natural pavement that stretches out into the sea, resembling a giant's stepping stones.

The formation of Giant's Causeway is a fascinating example of the interaction between volcanic activity and geological processes. Here is how the causeway came to be:

Volcanic Eruption: Around 60 million years ago, during the Paleogene period, the region experienced intense volcanic activity. Molten basalt lava erupted from fissures in the Earth's crust, flowing across the landscape and into the sea.

Cooling and Contraction: As the lava cooled, it began to contract and crack. The cooling process was relatively rapid, causing the basalt to fracture in a distinctive hexagonal pattern. This type of jointing, known as columnar jointing, is a common feature in basalt formations but is exceptionally well-developed and extensive at Giant's Causeway.

Erosion and Exposure: Over millions of years, erosion by wind, rain, and waves gradually exposed the basalt columns, creating the dramatic landscape visible today. The relentless action of the sea continues to shape and sculpt the causeway, revealing more of its intricate structure.

Giant's Causeway has inspired numerous myths and legends, the most famous of which involves the Irish giant Fionn mac Cumhaill (Finn McCool). According to legend, Fionn built the causeway to cross the sea to Scotland and confront his rival, the Scottish giant Benandonner. Upon realizing that Benandonner was much larger than himself, Fionn's wife disguised him as a baby. When Benandonner saw the "baby," he assumed that its father, Fionn, must be enormous, and fled back to Scotland, tearing up the causeway behind him. This tale adds a layer of cultural richness to the natural wonder, blending geology with folklore.

Visitors to Giant's Causeway can explore the unique landscape through a network of walking trails and viewpoints. The most famous features of the site include the "Wishing Chair," a natural formation resembling a throne, and the "Organ," a series of tall columns resembling organ pipes. The visitor center provides educational exhibits about the geological history, wildlife, and cultural significance of the causeway.

The biodiversity of Giant's Causeway is also noteworthy. The rock pools and cliffs provide habitats for a variety of marine life, seabirds, and plant species. The area's unique geology creates microhabitats that support rare and specialized species, contributing to its ecological value.

In addition to its natural beauty and geological significance, Giant's Causeway has played a role in scientific research. The formation has been studied extensively to understand the processes of columnar jointing and volcanic activity. It serves as a natural laboratory for geologists and volcanologists, offering insights into Earth's geological history and the dynamics of volcanic landscapes.

Conservation efforts are crucial to preserving Giant's Causeway. The site faces threats from erosion, climate change, and human activity. Managing visitor access, monitoring environmental impacts, and promoting sustainable tourism practices are essential to protect this geological treasure for future generations.

In conclusion, Giant's Causeway is a breathtaking example of the interplay between volcanic activity and geological processes. Its unique hexagonal basalt columns, formed by cooling and cracking lava, create a natural wonder that has captured the imagination of people for centuries. The site's blend of natural beauty, cultural legends, and scientific significance makes it a must-visit destination and a valuable resource for understanding Earth's dynamic geological history.

Should we modify this page or shall we move on to the next

2. Antelope Canyon

Antelope Canyon, located in the American Southwest near Page, Arizona, is a stunning geological formation renowned for its wave-like, sculpted walls and narrow passageways. This slot canyon, carved by erosion over thousands of years, is divided into two sections: Upper Antelope Canyon (or "The Crack") and Lower Antelope Canyon (or "The Corkscrew"). The intricate patterns, vibrant colors, and striking light beams make Antelope Canyon a popular destination for photographers, hikers, and nature enthusiasts.

The formation of Antelope Canyon is primarily the result of two natural forces: water erosion and wind. The process can be summarized as follows:

Sandstone Composition: Antelope Canyon is composed of Navajo Sandstone, a type of sedimentary rock that formed approximately 190 million years ago during the Jurassic period. This sandstone is relatively soft and porous, making it susceptible to erosion.

Flash Floods: The primary force behind the creation of Antelope Canyon is water erosion, particularly from flash floods. During periods of heavy rainfall, water flows rapidly through the narrow canyon, carrying with it sand, gravel, and other debris. This fast-moving water acts like sandpaper, carving away at the sandstone walls and creating the smooth, undulating surfaces and narrow passageways characteristic of slot canyons.

Wind Erosion: In addition to water, wind plays a significant role in shaping the canyon. Strong winds can carry sand and other particles, further smoothing the rock surfaces and enhancing the intricate patterns within the canyon walls.

Light and Color: The unique lighting within Antelope Canyon adds to its breathtaking beauty. The narrow openings at the top of the canyon allow sunlight to filter in, creating dramatic light beams and illuminating the sandstone's vibrant hues of red, orange, pink, and purple. The interplay of light and shadow changes throughout the day, offering different visual experiences depending on the time of the visit.

Upper Antelope Canyon is more accessible and popular among tourists due to its wider entrance and relatively flat terrain. The canyon's famous light beams are best observed during the summer months, around midday, when the sun is directly overhead. Lower Antelope Canyon, while narrower and requiring some climbing, offers equally stunning views and tends to be less crowded, providing a more intimate experience of the canyon's beauty.

Both sections of Antelope Canyon are located on Navajo land, and guided tours led by Navajo guides are required to visit the canyon. These tours not only provide access to the site but also offer valuable insights into the geological history, cultural significance, and safety considerations associated with exploring the canyon.

The cultural significance of Antelope Canyon extends beyond its natural beauty. For the Navajo people, the canyon is a sacred site,

193

deeply connected to their spiritual beliefs and traditions. The Navajo name for Upper Antelope Canyon is "Tsé bighánílíní," which means "the place where water runs through rocks." Lower Antelope Canyon is known as "Hazdistazí," meaning "spiral rock arches." Respect for the land and the stories associated with it is an integral part of visiting Antelope Canyon.

Safety is a critical concern when visiting slot canyons like Antelope Canyon. Flash floods can occur with little warning, even when there is no rain in the immediate area, as storms upstream can send torrents of water through the canyon. This makes it essential to visit with knowledgeable guides who monitor weather conditions and ensure the safety of visitors.

In addition to its cultural and aesthetic value, Antelope Canyon serves as a natural laboratory for studying the processes of erosion and sedimentation. Geologists and earth scientists study the canyon to understand the dynamics of flash floods, the properties of sandstone, and the long-term evolution of landscapes shaped by water and wind.

In conclusion, Antelope Canyon is a geological marvel shaped by the forces of water and wind over millennia. Its sculpted sandstone walls, vibrant colors, and dramatic light beams make it a must-visit destination for nature lovers and photographers. As a site of cultural significance to the Navajo people and a natural wonder, Antelope Canyon offers a unique and awe-inspiring experience, highlighting the beauty and power of the natural world.

3. The Wave

The Wave, located in the Paria Canyon-Vermilion Cliffs Wilderness area on the Arizona-Utah border, is one of the most visually striking and geologically fascinating formations in the American Southwest. Known for its undulating, wave-like patterns and vibrant colors, The Wave is a sandstone rock formation that has captivated photographers, hikers, and geologists alike.

The formation of The Wave is a result of millions of years of geological processes, primarily involving sedimentation, compaction, and erosion. Here's how this remarkable landscape came to be:

Sedimentation: The Wave is part of the Navajo Sandstone, which formed during the Jurassic period, around 190 million years ago. At that time, the region was covered by vast deserts with large sand dunes. Over time, layers of sand were deposited, creating cross-bedded layers that are characteristic of dune environments.

Compaction and Cementation: Over millions of years, the layers of sand were buried by additional sediments, leading to compaction. Groundwater rich in minerals flowed through the sand layers, cementing the grains together and forming solid sandstone. The different mineral compositions in the groundwater contributed to the varied colors seen in the rock today, with shades of red, orange, yellow, and pink.

Erosion: The unique wave-like patterns of The Wave were primarily created by erosion, driven by wind and water. The relatively soft sandstone was gradually worn away, with the wind playing a significant role in shaping the smooth, flowing lines and ridges. The cross-bedding of the original sand dunes is preserved in the rock, creating the stunning striations that give The Wave its distinctive appearance.

The Wave is a protected area, and access is strictly regulated to preserve its delicate and fragile landscape. A permit system is in place, with a limited number of daily permits issued through a lottery. This helps to minimize human impact and ensures that visitors can experience the beauty of The Wave without overcrowding.

Visitors to The Wave often describe it as a surreal and otherworldly experience. The formation's undulating surfaces and vibrant colors create a mesmerizing and immersive environment. Photographers, in particular, are drawn to The Wave for its unique light and shadow play, which changes throughout the day as the sun moves across the sky.

In addition to its aesthetic appeal, The Wave offers valuable insights into the geological history of the region. Geologists study the formation to understand the processes of sedimentation, compaction, and erosion, as well as the broader geological history of the Colorado Plateau. The cross-bedded sandstone layers provide a glimpse into the ancient desert environments that once covered the area.

The surrounding Paria Canyon-Vermilion Cliffs Wilderness is also home to a diverse array of plant and animal life. Visitors may encounter species such as desert bighorn sheep, pronghorn antelope, and various birds and reptiles. The wilderness area's diverse ecosystems and striking landscapes make it a destination for nature enthusiasts and adventure seekers.

While The Wave is undoubtedly the most famous formation in the area, the region is also home to other geological wonders, such as the Coyote Buttes, Buckskin Gulch, and the Vermilion Cliffs. Each of these locations offers its own unique geological features and opportunities for exploration and discovery.

Conservation efforts are critical to preserving The Wave and the surrounding wilderness. Visitors are encouraged to follow Leave No Trace principles, stay on designated trails, and respect the fragile environment. These measures help to ensure that future generations can continue to experience and appreciate this natural wonder.

In conclusion, The Wave is a breathtaking example of the power and beauty of geological processes. Its undulating sandstone formations, vibrant colors, and intricate patterns make it a must-visit destination for nature lovers, photographers, and geologists. As a protected and cherished natural wonder, The Wave offers a unique and immersive experience that highlights the dynamic and ever-changing nature of our planet's landscapes.

4. Mount Roraima

Mount Roraima, located at the triple border point of Venezuela, Brazil, and Guyana, is one of the most remarkable geological formations on Earth. This tepui, or tabletop mountain, is part of the Pakaraima Mountain Range and is known for its sheer cliffs, flat summit, and unique ecosystem. Mount Roraima has inspired myths, scientific exploration, and popular culture, making it a symbol of natural beauty and geological wonder.

The formation of Mount Roraima can be traced back to ancient geological processes that began over two billion years ago:

Precambrian Shield: The rocks that make up Mount Roraima are part of the Guiana Shield, one of the oldest geological formations on Earth. The Guiana Shield consists of Precambrian rock formations that have been stable for billions of years. Over time, layers of sandstone and quartzite were deposited on top of these ancient rocks.

Erosion and Weathering: Around 200 million years ago, during the Jurassic period, the region experienced significant geological uplift, forming the Pakaraima Mountains. Over millions of years, erosion and weathering processes sculpted the landscape, creating the characteristic flat-topped mesas known as tepuis. Mount Roraima's flat summit and steep cliffs are the result of these long-term erosional processes.

Unique Ecosystem: The isolated and elevated summit of Mount Roraima has created a unique ecosystem, home to numerous endemic species of plants and animals. The harsh and isolated

197

environment has allowed for the evolution of species that are found nowhere else on Earth. The summit is covered with rocky outcrops, small ponds, and unique vegetation adapted to the challenging conditions.

Mount Roraima's striking appearance and unique environment have inspired numerous myths and legends among the indigenous peoples of the region. The Pemon and Kapon tribes of Venezuela and Guyana regard the mountain as sacred and refer to it as the "House of the Gods." According to their mythology, Mount Roraima is the stump of a mighty tree that once held all the fruits and vegetables of the world, cut down by the trickster Makunaima. The mountain has also captured the imagination of explorers and writers. Sir Arthur Conan Doyle's 1912 novel "The Lost World" was inspired by accounts of Mount Roraima, depicting it as a place where prehistoric creatures still roam. This association with mystery and adventure has made Mount Roraima a symbol of the unknown and unexplored.

Mount Roraima is a popular destination for trekkers and adventurers seeking to experience its unique landscape and challenging ascent. The journey to the summit typically begins in the Gran Sabana region of Venezuela, involving a multi-day trek through diverse terrain, including rainforests, rivers, and steep ascents. The climb itself is physically demanding, but the rewards are unparalleled: panoramic views, surreal rock formations, and the sense of standing on one of the oldest geological formations on Earth.

The summit of Mount Roraima is a plateau approximately 31 square kilometers (12 square miles) in area, characterized by a harsh and otherworldly landscape. The constant mist and frequent rains create a unique microclimate, with temperatures often dropping below freezing at night. Despite these harsh conditions, the summit hosts a variety of plant species, including carnivorous plants like

Heliamphora, and unique animal species such as the Roraima black frog (Oreophrynella quelchii).

Scientific research on Mount Roraima has provided valuable insights into evolutionary biology, geology, and ecology. The isolation of the summit has led to the development of unique species and ecosystems, offering a natural laboratory for studying evolutionary processes. The geological formations also provide a window into Earth's ancient history, helping scientists understand the processes that have shaped our planet over billions of years.

In addition to its scientific and adventurous appeal, Mount Roraima holds cultural and symbolic significance. It represents a connection to ancient geological processes, indigenous cultures, and the human spirit of exploration and discovery. As a protected area within Canaima National Park in Venezuela, efforts are ongoing to preserve its unique environment and ensure that it remains a place of wonder for future generations.

In conclusion, Mount Roraima is a breathtaking and iconic geological formation that exemplifies the beauty and complexity of our planet's natural history. Its sheer cliffs, flat summit, and unique ecosystem make it a destination for adventurers, scientists, and those seeking a connection to the ancient past. As a symbol of mystery and exploration, Mount Roraima continues to inspire and captivate all who encounter its majestic presence.

5. Pamukkale

Pamukkale, located in southwestern Turkey, is a natural wonder renowned for its dazzling white terraces and thermal waters. The name "Pamukkale" translates to "Cotton Castle" in Turkish, aptly describing the cotton-like appearance of the travertine terraces. This UNESCO World Heritage Site is not only a geological marvel but also a significant historical and cultural landmark, attracting visitors from around the world.

The formation of Pamukkale's terraces and pools involves a combination of geological and hydrological processes:
Thermal Springs: Pamukkale is situated in the Menderes River valley, an area known for its geothermal activity. Underground volcanic activity heats the water, which becomes rich in minerals, particularly calcium carbonate, as it rises to the surface through fissures in the rock.
Calcium Carbonate Deposition: As the hot, mineral-rich water flows over the edge of the terraces, it cools and loses some of its carbon dioxide content. This reduction in carbon dioxide causes the calcium carbonate to precipitate out of the water, forming travertine deposits. Over time, these deposits build up to create the terraces, which are characterized by their brilliant white color.
Terrace Formation: The travertine terraces are formed by a series of cascading pools that step down the hillside. The water flows from one pool to the next, depositing more calcium carbonate and continually reshaping the terraces. The resulting formations resemble a series of frozen waterfalls or a natural amphitheater made of white stone.
Pamukkale has been a popular destination for millennia, dating back to the ancient Greco-Roman city of Hierapolis, which was founded around 190 BCE. The city's inhabitants were drawn to the area by the thermal springs, which they believed had healing properties. Hierapolis became a major spa center, attracting visitors seeking treatment for various ailments. The ruins of Hierapolis, which include a theater, temples, baths, and a necropolis, are situated adjacent to Pamukkale and add to the site's historical significance.
Today, Pamukkale and Hierapolis are jointly recognized as a UNESCO World Heritage Site, attracting millions of tourists each year. Visitors can walk barefoot along designated paths on the terraces, allowing them to experience the unique texture of the

travertine and the warm, mineral-rich waters. The thermal pools provide a relaxing and therapeutic experience, continuing the ancient tradition of using Pamukkale's waters for their purported health benefits.

The preservation of Pamukkale's delicate travertine terraces is a priority for local authorities and conservationists. In the past, unregulated tourism and the construction of hotels and roads near the terraces led to damage and discoloration of the travertine. Efforts have since been made to protect and restore the site, including the removal of modern structures, the restriction of access to certain areas, and the implementation of water management practices to ensure a sustainable flow of thermal water.

Pamukkale's unique geological and hydrological features make it an important site for scientific research. Geologists and hydrologists study the formation and dynamics of the travertine terraces to understand the processes of mineral deposition, thermal spring activity, and landscape evolution. The site also provides insights into the relationship between geothermal activity and the Earth's crust, contributing to the broader field of geoscience.

In addition to its geological and historical significance, Pamukkale holds cultural and spiritual importance. The thermal waters and travertine terraces have been revered for centuries, and the site continues to inspire awe and wonder in those who visit. The combination of natural beauty, historical ruins, and therapeutic waters creates a unique and enriching experience for travelers.

Pamukkale also serves as a reminder of the delicate balance between human activity and natural preservation. The efforts to protect and maintain the site highlight the importance of sustainable tourism and environmental stewardship. By preserving Pamukkale, we ensure that future generations can continue to marvel at this extraordinary natural wonder and appreciate its historical and cultural legacy.

In conclusion, Pamukkale is a stunning geological formation that captivates with its white travertine terraces and thermal waters. Formed by the deposition of calcium carbonate from mineral-rich springs, the terraces have been a site of human interest for thousands of years. As a UNESCO World Heritage Site, Pamukkale offers a unique blend of natural beauty, historical significance, and therapeutic value, making it a must-visit destination for those seeking to experience one of the world's most remarkable geological wonders.

6. The Marble Caves

The Marble Caves, also known as the Marble Cathedral or Marble Chapels (Capillas de Mármol in Spanish), are one of the most stunning natural wonders in Patagonia, Chile. Located on the General Carrera Lake (Lago General Carrera), these mesmerizing caves are renowned for their vibrant colors and intricate, swirling patterns of marble, sculpted over thousands of years by water erosion.

The formation of the Marble Caves involves a combination of geological and hydrological processes:

Limestone Deposition: The marble formations in the caves began as limestone deposits formed over millions of years from the accumulation of marine organisms. These deposits were subjected to intense pressure and heat from tectonic activity, transforming the limestone into crystalline marble.

Water Erosion: The Marble Caves owe their intricate shapes and vibrant colors to the erosive power of the waters of General Carrera Lake. Over thousands of years, the lake's waves have gradually eroded the marble, carving out intricate caverns, tunnels, and pillars. The constant movement of the water has smoothed the surfaces of the marble, creating the unique swirling patterns and undulating formations seen today.

Color Variations: The striking colors of the Marble Caves, ranging from bright turquoise to deep blue, are a result of the interaction between the crystalline marble and the clear waters of the lake. The color of the water, influenced by the mineral content and light conditions, enhances the natural hues of the marble, creating a mesmerizing visual effect.

Visitors to the Marble Caves can explore the formations by boat or kayak, offering an up-close view of the stunning geological features. The best time to visit is during the summer months (December to March) when the water levels are lower, revealing more of the caves' intricate structures. The sunlight during this period enhances the colors of the water and the marble, creating a spectacular display of natural beauty.

The main attractions within the Marble Caves complex include the Marble Cathedral (Catedral de Mármol) and the Marble Chapel (Capilla de Mármol). These formations are particularly renowned for their grandeur and the complexity of their shapes, resembling the architectural intricacies of a cathedral or chapel.

The Marble Caves are not only a visual marvel but also a significant geological site. The process of marble formation and erosion provides valuable insights into the geological history and natural processes of the region. Studying these formations helps scientists understand the long-term effects of water erosion on carbonate rocks and the role of tectonic activity in shaping the landscape.

In addition to their geological importance, the Marble Caves are a key tourist attraction in Patagonia, drawing visitors from around the world. The site offers a unique blend of adventure and natural beauty, with opportunities for boating, kayaking, and photography. The remote location of the Marble Caves adds to their allure, providing an off-the-beaten-path experience for travelers seeking to explore one of nature's hidden gems.

Conservation efforts are essential to preserving the Marble Caves and their pristine environment. The delicate nature of the marble formations and the surrounding ecosystem requires careful management to ensure that tourism does not negatively impact the site. Sustainable tourism practices, such as limiting the number of visitors and promoting responsible behavior, are crucial for maintaining the integrity of this natural wonder.

The Marble Caves also hold cultural significance for the local communities. The region around General Carrera Lake is home to a rich cultural heritage, with indigenous and local traditions intertwined with the natural landscape. The caves are a source of pride and inspiration, reflecting the beauty and resilience of the Patagonian environment.

In conclusion, the Marble Caves of Patagonia, Chile, are a breathtaking example of nature's artistry, carved and colored by the erosive power of water over thousands of years. These intricate marble formations, with their vibrant hues and swirling patterns, offer a unique and awe-inspiring experience for visitors. As a site of geological, cultural, and aesthetic significance, the Marble Caves exemplify the dynamic and ever-changing beauty of our planet's natural landscapes, highlighting the importance of conservation and sustainable tourism to protect such precious natural treasures.

7. The Eye of the Sahara (Richat Structure)

The Eye of the Sahara, also known as the Richat Structure, is a prominent and enigmatic geological formation located in the Sahara Desert near Ouadane, Mauritania. This circular structure, approximately 40 kilometers (25 miles) in diameter, resembles a giant bullseye or an eye, making it easily recognizable from space.

The Richat Structure has intrigued scientists and geologists for decades, sparking various theories about its origin and significance. The formation of the Eye of the Sahara can be attributed to a combination of geological processes, including volcanic activity, erosion, and uplift. Here is an overview of how this unique structure came to be:

Volcanic Activity and Uplift: The Richat Structure is believed to have formed through volcanic activity during the Proterozoic Eon, around 2 billion years ago. Magma from deep within the Earth's crust pushed upwards, creating a domed structure. This process, known as igneous intrusion, resulted in the uplift of the overlying rock layers.

Erosion: Over millions of years, the raised dome was subjected to extensive erosion by wind and water. The outer layers of rock were gradually worn away, revealing the concentric rings of different rock types that make up the Richat Structure. These rings consist of various sedimentary rocks, including sandstone, limestone, and dolomite, which have been exposed in a circular pattern.

Differential Weathering: The varying resistance of the rock layers to erosion has contributed to the distinctive appearance of the Eye of the Sahara. Harder, more resistant rock layers form the prominent ridges, while softer, less resistant layers have eroded more quickly, creating the depressions between the ridges. This differential weathering has resulted in the concentric rings that are visible today.

Initially, the Eye of the Sahara was thought to be an impact crater formed by a meteorite collision. However, geological studies have since ruled out this hypothesis, as there is no evidence of shock metamorphism or other indicators typically associated with impact craters. Instead, the current understanding is that the structure is a deeply eroded, uplifted dome.

The Richat Structure is not only a geological curiosity but also a valuable site for scientific research. The exposed rock layers provide insights into the geological history of the region, including the types of sedimentary environments that existed billions of years ago. Studying the Richat Structure helps geologists understand the processes of volcanic activity, erosion, and weathering, as well as the tectonic history of the Sahara Desert.

The Eye of the Sahara is also a significant landmark for space exploration and observation. Its distinct shape and size make it a useful reference point for astronauts and satellites. The structure has been photographed extensively from space, contributing to its recognition and study.

In addition to its scientific importance, the Eye of the Sahara holds cultural and historical significance. The region surrounding the structure has been inhabited by various human civilizations over thousands of years. Archaeological evidence suggests that ancient peoples used the area for settlement and trade, leaving behind artifacts and remnants of their presence. The Richat Structure serves as a reminder of the long history of human interaction with the environment.

The Eye of the Sahara's remote location and harsh desert conditions make it a challenging destination for tourists. However, for those who venture to the site, the experience is unforgettable. Visitors can explore the concentric rings, hike the rugged terrain, and witness the vastness of the Sahara Desert from a unique vantage point. The site's natural beauty and geological intrigue attract adventurers, geologists, and photographers alike.

Conservation efforts are essential to protect the Richat Structure and its surrounding environment. The fragile desert ecosystem is vulnerable to the impacts of human activity, including tourism and development. Sustainable practices and responsible tourism are crucial to preserving the site's integrity and ensuring that it remains a valuable resource for scientific research and natural beauty.

In conclusion, the Eye of the Sahara, or the Richat Structure, is a fascinating geological formation that captivates with its circular, bullseye-like appearance. Formed through volcanic activity, uplift, and erosion, this unique structure offers valuable insights into geological processes and the history of the Sahara Desert. As a site of scientific, cultural, and natural significance, the Eye of the Sahara continues to inspire curiosity and exploration, highlighting the intricate and dynamic nature of our planet's geological history.

8. The Door to Hell (Darvaza Gas Crater)

The Darvaza Gas Crater, commonly known as the "Door to Hell," is a captivating and eerie geological phenomenon located in the Karakum Desert of Turkmenistan. This burning crater, measuring about 70 meters (230 feet) in diameter and 20 meters (66 feet) deep, has been continuously aflame since its accidental creation in 1971. The intense heat, glowing flames, and boiling mud within the crater give it an otherworldly appearance, drawing adventurous travelers and scientists alike.

The creation of the Darvaza Gas Crater was an unintended consequence of Soviet-era natural gas drilling. Here's how this geological marvel came to be:

Natural Gas Exploration: In 1971, Soviet geologists were drilling for natural gas in the Karakum Desert, a region known for its rich reserves of oil and natural gas. During the drilling process, they encountered a natural gas cavern.

Collapse and Formation: The drilling rig accidentally punctured the cavern, causing the ground to collapse and forming a large crater. The collapse released a significant amount of methane gas into the atmosphere, posing a potential hazard.

Ignition: To prevent the spread of methane gas, which is highly flammable and a potent greenhouse gas, the geologists decided to ignite the gas, expecting it to burn off within a few weeks. However, the gas has continued to burn for decades, creating the perpetual flames that have earned the site its nickname, the "Door to Hell."

The continuous burning of the Darvaza Gas Crater is fueled by the natural gas reserves beneath the surface. The flames, which can reach heights of several meters, produce a hellish glow that is visible from miles away, especially at night. The intense heat emitted by the crater can be felt even from a distance, and the sight of the fiery pit amidst the vast desert landscape is both mesmerizing and unsettling.

The Darvaza Gas Crater has become a unique tourist attraction, drawing visitors from around the world who are intrigued by its unusual and dramatic appearance. Adventurous travelers often camp near the crater to witness the fiery spectacle, particularly at night when the glow is most pronounced. Despite its remote location, the crater's notoriety has made it a must-see destination for those exploring Turkmenistan and the Karakum Desert.

The site also holds scientific interest due to its unique environmental conditions. Researchers study the Darvaza Gas Crater to understand the geological and chemical processes at play. The constant burning of methane provides insights into the behavior of natural gas reserves and the impact of such events on the local environment. Additionally, the extreme conditions within the crater offer a natural laboratory for studying extremophiles—organisms that thrive in extreme environments.

The Darvaza Gas Crater has also raised discussions about environmental concerns and the management of natural gas resources. The continuous release of methane, even in a burning state, contributes to greenhouse gas emissions. While the flames

208

prevent the uncontrolled release of methane into the atmosphere, the burning still produces carbon dioxide. This highlights the need for improved practices in natural gas extraction and management to minimize environmental impacts.

Efforts to address the environmental implications of the Darvaza Gas Crater have been limited due to its unique nature and the challenges associated with extinguishing the flames. Some proposals have suggested ways to harness the burning gas for energy production, while others emphasize the importance of preserving the site as a natural and geological curiosity. Balancing the environmental considerations with the site's value as a tourist attraction and research site remains a complex issue.

In conclusion, the Darvaza Gas Crater, or the "Door to Hell," is a striking example of the unintended consequences of natural gas exploration. Its perpetual flames and hellish appearance create a fascinating and eerie spectacle in the Karakum Desert. As a site of scientific interest, environmental concern, and adventurous tourism, the Darvaza Gas Crater highlights the intricate and sometimes unpredictable interactions between human activity and natural geological processes. Preserving and studying this unique phenomenon offers valuable insights into both the beauty and challenges of our planet's dynamic landscapes.

9.   The Chocolate Hills

The Chocolate Hills, located in the Bohol province of the Philippines, are a unique and visually striking geological formation consisting of more than 1,200 conical hills spread over an area of about 50 square kilometers (19 square miles). These hills, which turn brown during the dry season, resemble giant chocolate mounds, hence their name. The Chocolate Hills are not only a natural wonder but also a national geological monument and a symbol of Bohol, attracting tourists and researchers from around the world.

The formation of the Chocolate Hills is the result of complex geological processes that began millions of years ago:

Limestone Deposition: During the Late Pliocene to Early Pleistocene period, the area that is now Bohol was submerged under the sea. Marine organisms, such as corals and mollusks, accumulated on the seafloor, forming thick layers of limestone. Over time, these limestone deposits were uplifted due to tectonic activity, creating a landscape of rolling hills and valleys.

Weathering and Erosion: The unique conical shape of the Chocolate Hills is primarily due to weathering and erosion. Rainwater, which is slightly acidic, gradually dissolved the limestone, carving out depressions and leaving behind the more resistant, cone-shaped hills. This process, known as karstification, is responsible for the creation of other karst landscapes around the world, but the Chocolate Hills are particularly notable for their uniformity in shape and size.

Vegetation and Color Change: The hills are covered in grass, which turns brown during the dry season, giving the hills their chocolate-like appearance. During the wet season, the grass is lush and green, transforming the landscape into a verdant sea of mounds.

The Chocolate Hills are a significant tourist attraction, drawing visitors who are fascinated by their unusual appearance and geological origins. Several viewing stations and observation decks have been established to provide panoramic views of the hills, with the most popular being the Chocolate Hills Complex in Carmen and the Sagbayan Peak.

In addition to their natural beauty, the Chocolate Hills hold cultural and historical significance. They are featured in local legends and folklore, with one popular tale describing the hills as the result of a giant's tears. According to the legend, the giant Arogo cried uncontrollably after losing his beloved, and his tears solidified into the hills we see today. Another story tells of two feuding giants who hurled boulders at each other, and their conflict eventually created the hills.

The Chocolate Hills also play a role in the local economy, with tourism providing a source of income for the communities in Bohol. Efforts are made to promote sustainable tourism and protect the natural environment to ensure that the hills remain a valuable resource for future generations.

Scientific research on the Chocolate Hills has provided valuable insights into karst landscapes and the geological history of the Philippines. Geologists study the hills to understand the processes of limestone formation, uplift, and erosion, as well as the broader tectonic activity in the region. The uniformity and density of the hills make them an interesting subject for comparative studies with other karst formations around the world.

Conservation efforts are essential to protect the Chocolate Hills from environmental threats, such as deforestation, quarrying, and climate change. Initiatives to preserve the hills include reforestation projects, sustainable land management practices, and the establishment of protected areas. These efforts aim to maintain the ecological integrity and natural beauty of the Chocolate Hills while supporting the livelihoods of local communities.

In conclusion, the Chocolate Hills of Bohol, Philippines, are a remarkable geological formation that captivates with their uniform conical shapes and seasonal color changes. Formed through millions of years of limestone deposition, weathering, and erosion, these hills are a testament to the power of natural geological processes. As a site of natural beauty, cultural significance, and scientific interest, the Chocolate Hills continue to inspire wonder and admiration, highlighting the intricate and dynamic nature of our planet's landscapes.

10. The Valley of the Moon (Wadi Rum)

Wadi Rum, also known as the Valley of the Moon, is a spectacular desert landscape located in southern Jordan. This UNESCO World Heritage site is renowned for its dramatic rock formations, towering sandstone cliffs, and expansive red sand dunes. The unique geological and cultural significance of Wadi Rum has made it a popular destination for tourists, filmmakers, and researchers.

The formation of Wadi Rum's striking landscape is the result of millions of years of geological processes:

Tectonic Activity: The region that is now Wadi Rum was influenced by tectonic forces that created faults and fractures in the Earth's crust. These tectonic movements caused the uplift and tilting of the sandstone layers, forming the dramatic cliffs and mountains that dominate the landscape.

Erosion: Over millions of years, wind and water erosion sculpted the sandstone into the stunning shapes seen today. The constant action of windblown sand has carved intricate patterns, arches, and narrow canyons, while occasional flash floods have further shaped the terrain.

Weathering: The extreme temperature fluctuations in the desert environment contribute to the weathering of the rock. The intense heat during the day and the cold temperatures at night cause the rock to expand and contract, leading to cracks and the gradual breakdown of the sandstone.

Wadi Rum's geological features include some of the most iconic and visually striking formations in the desert:

The Seven Pillars of Wisdom: Named after T.E. Lawrence's book, these seven towering rock formations are one of Wadi Rum's most recognizable landmarks. The pillars rise majestically from the desert floor, creating a dramatic and picturesque scene.

Khazali Canyon: This narrow gorge is famous for its ancient rock inscriptions and petroglyphs, which date back thousands of years.

212

The walls of the canyon are adorned with carvings depicting humans, animals, and symbols, providing a glimpse into the region's rich cultural history.

Um Frouth Rock Bridge: One of several natural rock bridges in Wadi Rum, Um Frouth is a popular spot for visitors to climb and take in the breathtaking views of the desert landscape. The bridge was formed by wind and water erosion, creating a stunning natural arch.

Burdah Rock Bridge: This is the largest rock arch in Wadi Rum and one of the largest in the world. The challenging hike to the top of the arch rewards visitors with panoramic views of the desert and surrounding mountains.

Wadi Rum is not only a geological wonder but also a place of great cultural and historical significance. The area has been inhabited by various cultures for millennia, including the Nabataeans, who left behind numerous rock carvings and temples. The Bedouin people have lived in Wadi Rum for generations, and their rich traditions and hospitality add to the cultural experience of visiting the valley.

The stunning landscape of Wadi Rum has made it a popular location for filmmakers, serving as a backdrop for numerous movies, including "Lawrence of Arabia," "The Martian," and "Star Wars: The Rise of Skywalker." The valley's otherworldly scenery makes it an ideal stand-in for alien planets and distant worlds.

Tourism in Wadi Rum offers a variety of activities for visitors, including jeep tours, camel rides, hiking, rock climbing, and stargazing. The clear desert skies and minimal light pollution make Wadi Rum an excellent location for observing the night sky, with opportunities to see constellations, planets, and even the Milky Way.

Conservation efforts are crucial to preserving the unique environment of Wadi Rum. The increasing number of visitors poses challenges to the fragile desert ecosystem. Sustainable

213

tourism practices, such as limiting the number of vehicles, reducing waste, and promoting eco-friendly accommodations, are essential to protect the landscape and ensure that it remains pristine for future generations.

Wadi Rum also plays an important role in scientific research. Geologists study the rock formations to understand the processes of erosion, weathering, and tectonic activity. The desert's unique flora and fauna, adapted to the harsh environment, provide insights into biodiversity and adaptation strategies. Archaeologists and historians explore the ancient inscriptions and artifacts to learn more about the region's past civilizations and cultural heritage.

In conclusion, Wadi Rum, the Valley of the Moon, is a breathtaking and iconic geological formation that showcases the power of natural processes over millions of years. Its towering cliffs, intricate rock formations, and expansive sand dunes create a landscape of unparalleled beauty and significance. As a site of cultural, historical, and scientific interest, Wadi Rum continues to inspire awe and wonder, highlighting the dynamic and ever-changing nature of our planet's geological history. Sustainable tourism and conservation efforts are vital to preserving this extraordinary natural wonder for future generations to experience and appreciate.

Chapter 10: Bizarre Plants and Fungi

1. Rafflesia arnoldii

Rafflesia arnoldii, commonly known as the corpse flower, is one of the most extraordinary and bizarre plants on Earth. Native to the rainforests of Southeast Asia, particularly in Indonesia and Malaysia, this parasitic plant is renowned for producing the largest individual flower in the world. The flower can reach up to 1 meter (3.3 feet) in diameter and weigh up to 11 kilograms (24 pounds). Despite its impressive size and striking appearance, Rafflesia arnoldii is perhaps best known for its pungent odor, which resembles that of rotting flesh.

The life cycle and characteristics of Rafflesia arnoldii are as unique as they are fascinating:

Parasitic Nature: Rafflesia arnoldii is a holoparasitic plant, meaning it lacks chlorophyll and is entirely dependent on its host for nutrients. The plant's seeds infiltrate the roots or stems of a specific host vine, Tetrastigma, which belongs to the grape family. The seeds germinate within the host tissue, forming an intricate network of thread-like filaments that absorb water and nutrients.

Hidden Growth: For most of its life cycle, Rafflesia arnoldii remains hidden within its host. The parasitic filaments grow and develop inside the host vine, without any visible signs of the plant above ground. This period of dormancy can last several months to years, during which the plant undergoes significant growth and development.

Emergence and Blooming: When Rafflesia arnoldii is ready to reproduce, it emerges from the host vine as a large, bud-like structure. Over the course of several days, the bud swells and eventually opens into a massive, five-petaled flower. The flower's color ranges from deep red to orange, with white, wart-like spots covering the petals.

Odor and Pollination: The flower emits a strong, foul odor that mimics the smell of decaying flesh. This odor serves to attract carrion flies and beetles, which are the primary pollinators of Rafflesia arnoldii. The insects are drawn to the flower, thinking it is a source of food. As they crawl over the flower searching for food, they inadvertently transfer pollen from one flower to another, facilitating cross-pollination.

Short Lifespan: The bloom of Rafflesia arnoldii is relatively short-lived, lasting only about five to seven days. After pollination, the flower begins to wilt and decay. If successfully pollinated, the flower develops into a fruit containing numerous seeds. These seeds are eventually dispersed by animals, such as small mammals, which consume the fruit and excrete the seeds in new locations, continuing the life cycle.

Rafflesia arnoldii is not only a marvel of nature due to its size and odor but also because of its elusive and enigmatic nature. The plant's reliance on a specific host vine and the difficulty of locating it in the dense rainforest make it a rare and prized find for botanists and nature enthusiasts.

The conservation status of Rafflesia arnoldii is a concern, as its habitat is threatened by deforestation, logging, and land conversion for agriculture. Efforts to protect and conserve the rainforests of Southeast Asia are crucial to preserving this unique plant species and its intricate ecological interactions.

In addition to its biological and ecological significance, Rafflesia arnoldii holds cultural importance in the regions where it is found. It is often considered a symbol of the rich biodiversity and natural heritage of Southeast Asia. The plant has been featured in local folklore, traditional medicine, and even as a national symbol in countries like Indonesia, where it is the official state flower of Sabah.

Scientific research on Rafflesia arnoldii continues to uncover new insights into its complex life cycle, genetics, and ecological relationships. Understanding the plant's parasitic mechanisms and its interactions with host vines and pollinators can contribute to broader knowledge in fields such as plant biology, ecology, and conservation science.

In conclusion, Rafflesia arnoldii, the corpse flower, stands out as one of the most bizarre and fascinating plants in the world. Its massive size, foul odor, and parasitic lifestyle make it a unique subject of study and admiration. As both a symbol of natural wonder and a reminder of the importance of conservation, Rafflesia arnoldii captures the imagination and highlights the incredible diversity of life on our planet.

2.   Hydnellum peckii

Hydnellum peckii, commonly known as the bleeding tooth fungus, is one of the most peculiar and visually striking fungi in the world. This fungus, belonging to the family Bankeraceae, is easily recognizable by its unusual appearance, which resembles a bleeding, tooth-like structure. Found primarily in North America and Europe, the bleeding tooth fungus is both a fascinating subject for mycologists and a curious sight for nature enthusiasts.

The distinctive features and life cycle of Hydnellum peckii are as follows:

217

Appearance: The bleeding tooth fungus gets its name from its distinctive fruiting body, which typically appears as a white to pale pink, cushion-like structure with tooth-like spines on its underside. When the fungus is young and actively growing, it exudes a bright red fluid that seeps out of tiny pores on the surface, giving it the appearance of a "bleeding" organism. This red fluid is actually a form of guttation, a process by which the fungus expels excess moisture and metabolic byproducts.

Habitat: Hydnellum peckii is typically found in coniferous forests, where it forms mycorrhizal associations with trees such as pines and spruces. These symbiotic relationships are beneficial for both the fungus and the host tree, as the fungus helps the tree absorb water and nutrients from the soil, while the tree provides the fungus with carbohydrates produced through photosynthesis.

Spore Dispersal: The spines on the underside of the fruiting body produce spores, which are released into the environment for reproduction. The spores are carried by wind, animals, and water, eventually landing in suitable habitats where they can germinate and form new fungal colonies.

Ecological Role: As a mycorrhizal fungus, Hydnellum peckii plays a crucial role in forest ecosystems. Its symbiotic relationship with trees enhances nutrient uptake and promotes healthy forest growth. Additionally, the fungus contributes to soil formation and stability by breaking down organic matter and recycling nutrients.

Chemical Compounds: The red fluid exuded by the bleeding tooth fungus contains various chemical compounds, including pigments and anticoagulants. These compounds have intrigued researchers, who study them for potential medicinal and industrial applications. Some compounds found in the fungus have shown antimicrobial properties, making them of interest for developing new antibiotics.

Despite its striking appearance, Hydnellum peckii is not considered edible. The fungus has a bitter taste and tough texture, making it unpalatable to humans. However, it poses no known threat to human health and is not considered toxic. The fungus's unique appearance and rarity make it a popular subject for photography and nature observation.

The bleeding tooth fungus is often found in undisturbed forest habitats, making it a good indicator of forest health and biodiversity. Its presence suggests a well-functioning ecosystem with healthy tree-fungal relationships. However, like many fungi, Hydnellum peckii is sensitive to environmental changes, such as deforestation, pollution, and climate change. Conservation efforts aimed at protecting forest habitats are essential to ensure the continued survival of this and other mycorrhizal fungi.

Hydnellum peckii also holds cultural and symbolic significance. Its eerie, blood-like appearance has earned it a place in folklore and popular culture, often associated with themes of mystery and the supernatural. The fungus has been featured in various media, including books, movies, and art, where its bizarre and otherworldly look captures the imagination.

Scientific research on Hydnellum peckii and other mycorrhizal fungi provides valuable insights into the complex interactions between fungi and plants. Understanding these relationships can inform conservation strategies, forest management practices, and the development of sustainable agriculture. Additionally, the study of fungal metabolites offers potential applications in medicine, biotechnology, and environmental remediation.

In conclusion, Hydnellum peckii, the bleeding tooth fungus, is a remarkable and bizarre organism that stands out for its unique appearance and ecological role. Its symbiotic relationships with trees, striking red guttation, and potential medicinal properties make it a fascinating subject for scientific research and nature enthusiasts. As a symbol of the intricate and diverse world of fungi,

Hydnellum peckii highlights the importance of conserving forest ecosystems and appreciating the wonders of the natural world.

3. Welwitschia mirabilis

Welwitschia mirabilis, often simply called Welwitschia, is one of the most bizarre and extraordinary plants on the planet. Native to the harsh deserts of Namibia and Angola in southwestern Africa, this ancient plant is renowned for its unique appearance and incredible resilience. Described as a "living fossil," Welwitschia has remained relatively unchanged for millions of years, providing a fascinating glimpse into the distant past of plant evolution.

The characteristics and life cycle of Welwitschia mirabilis are both unusual and remarkable:

Appearance: Welwitschia is easily recognizable by its distinctive morphology. The plant consists of only two leaves, a stem base, and a taproot. The two strap-like leaves are broad, leathery, and can grow up to 2-4 meters (6-13 feet) long. Unlike most plants, Welwitschia's leaves grow continuously from the base, gradually becoming tattered and split as they age. The stem base, or caudex, is thick and woody, resembling a squat, barrel-like structure that can reach up to 1 meter (3 feet) in height.

Longevity: One of the most remarkable aspects of Welwitschia is its longevity. Individual plants can live for hundreds, if not thousands, of years. Some estimates suggest that the oldest specimens may be over 1,500 years old. The plant's slow growth rate and ability to survive in extreme conditions contribute to its exceptional lifespan.

Adaptations: Welwitschia is highly adapted to its arid environment. The leaves' broad surface area allows them to capture moisture from fog and dew, which is then absorbed by specialized cells. The thick, waxy cuticle on the leaves reduces water loss through evaporation. The plant's deep taproot can reach groundwater sources, providing additional moisture during prolonged dry periods.

Reproduction: Welwitschia is dioecious, meaning that individual plants are either male or female. The reproductive structures are cone-like and located at the center of the plant. Male plants produce small, pollen-bearing cones, while female plants produce larger cones containing ovules. Pollination is primarily carried out by insects, particularly beetles, which are attracted to the nectar produced by the cones.

Ecological Role: In its native desert habitat, Welwitschia plays an important ecological role. The plant provides shelter and food for various desert-dwelling organisms, including insects, reptiles, and small mammals. Its presence indicates a stable, well-adapted desert ecosystem.

Welwitschia mirabilis holds significant scientific interest due to its unique evolutionary history and adaptations. The plant belongs to its own family, Welwitschiaceae, and is the sole surviving member of the order Welwitschiales. Genetic studies suggest that Welwitschia diverged from other seed plants around 100 million years ago, making it a critical species for understanding plant evolution and diversification.

In addition to its scientific value, Welwitschia has cultural and symbolic significance for the indigenous peoples of Namibia and Angola. The plant is often regarded with reverence and is sometimes used in traditional medicine and rituals. Its resilience and ability to thrive in one of the harshest environments on Earth make it a symbol of endurance and adaptability.

Conservation efforts are crucial to protect Welwitschia mirabilis and its habitat. The plant is classified as a vulnerable species due to threats such as habitat destruction, climate change, and over-collection by plant enthusiasts. Conservation measures include habitat protection, sustainable land use practices, and raising awareness about the importance of preserving this unique plant species.

Welwitschia has also captured the imagination of artists, writers, and photographers. Its otherworldly appearance and ancient lineage have made it a subject of fascination and inspiration. The plant's ability to survive and thrive in extreme conditions serves as a powerful reminder of nature's resilience and adaptability.

In conclusion, Welwitschia mirabilis is a truly bizarre and extraordinary plant that stands out for its unique appearance, longevity, and remarkable adaptations. As a living fossil and a symbol of resilience, Welwitschia provides valuable insights into plant evolution and the complexities of desert ecosystems. Conservation efforts are essential to ensure that this ancient and fascinating species continues to thrive in its natural habitat, offering a living link to the distant past of our planet's botanical history.

4. Amorphophallus titanum

Amorphophallus titanum, commonly known as the titan arum or corpse flower, is one of the most remarkable and bizarre plants in the world. Native to the rainforests of Sumatra, Indonesia, this gigantic flowering plant is famous for its enormous inflorescence and its pungent odor, which resembles that of rotting flesh. The titan arum is a member of the Araceae family and holds the record for producing the largest unbranched inflorescence in the plant kingdom.

The life cycle and characteristics of Amorphophallus titanum are fascinating and complex:

Structure and Size: The most striking feature of the titan arum is its massive inflorescence, which can reach heights of up to 3 meters (10 feet) and diameters of 1.5 meters (5 feet). The inflorescence consists of a central spike called the spadix, which is surrounded by a petal-like structure known as the spathe. The spathe is typically green on the outside and deep burgundy on the inside, giving the flower its characteristic appearance.

Odor and Pollination: When the titan arum blooms, it emits a strong, foul odor reminiscent of decaying flesh. This odor is produced by a combination of chemicals, including dimethyl trisulfide, which is also found in decomposing animal tissue. The stench attracts carrion beetles and flesh flies, which serve as the primary pollinators. These insects are lured by the smell, believing it to be a source of food or a suitable place to lay eggs. As they crawl over the flower, they inadvertently transfer pollen from one inflorescence to another, facilitating cross-pollination.

Blooming Cycle: The blooming of the titan arum is an infrequent and spectacular event. The plant undergoes a long period of vegetative growth, during which it produces a single, large leaf that can reach up to 6 meters (20 feet) in height and 5 meters (16 feet) in width. This leaf functions as a temporary tree, photosynthesizing and storing energy in the underground tuber. After several years of vegetative growth, the plant goes into a dormant phase before producing an inflorescence. The blooming period lasts only 24 to 48 hours, after which the spathe collapses and the spadix withers.

Tuber and Growth: The underground tuber of the titan arum is another remarkable feature. It can weigh up to 100 kilograms (220 pounds) and serves as the plant's energy storage organ. The tuber enables the plant to survive the long periods between blooms and supports the rapid growth of the inflorescence and leaf.

Conservation and Cultivation: Amorphophallus titanum is classified as vulnerable due to habitat loss and deforestation in its native Sumatra. Botanical gardens and research institutions around the world have successfully cultivated the titan arum, helping to raise awareness about its conservation and providing opportunities for scientific study. The blooming of a titan arum in cultivation is a major event, often attracting large crowds of visitors eager to witness the rare spectacle.

The titan arum holds significant cultural and scientific interest. Its extraordinary size, unusual life cycle, and pungent odor have made it a subject of fascination for botanists, horticulturists, and the general public. The plant has also inspired various works of art, literature, and media, symbolizing the wonders and mysteries of the natural world.

In addition to its aesthetic and cultural value, the titan arum provides important ecological functions in its native habitat. By attracting carrion beetles and flesh flies, the plant plays a role in the pollination and reproductive success of other plant species that rely on these insects for pollination.

Scientific research on Amorphophallus titanum continues to uncover new insights into plant physiology, reproductive biology, and chemical ecology. Understanding the mechanisms behind its rapid growth, chemical production, and energy storage can contribute to broader knowledge in plant science and inform conservation strategies for other endangered species.

In conclusion, Amorphophallus titanum, the titan arum or corpse flower, is one of the most extraordinary and bizarre plants on Earth. Its immense size, putrid odor, and infrequent blooming make it a unique and fascinating subject of study and admiration. As both a symbol of natural wonder and a reminder of the importance of conservation, the titan arum captures the imagination and highlights the incredible diversity of life in our planet's ecosystems.

5.   Clathrus archeri

Clathrus archeri, commonly known as the Devil's Fingers or octopus stinkhorn, is one of the most unusual and eerie fungi in the world. Native to Australia and Tasmania, this peculiar fungus has spread to other parts of the world, including Europe and North America. Its striking appearance, combined with its foul odor, makes Clathrus archeri a fascinating subject for mycologists and nature enthusiasts.

The life cycle and characteristics of Clathrus archeri are as bizarre as they are intriguing:

Appearance: The most distinctive feature of Clathrus archeri is its mature fruiting body, which resembles a cluster of bright red or orange tentacle-like fingers, reminiscent of an octopus or demonic hand. These "fingers" emerge from an egg-like structure that initially appears on the ground. The fingers are covered with a slimy, spore-bearing substance that emits a strong, unpleasant odor.

Lifecycle: The life cycle of Clathrus archeri begins with the development of a whitish, gelatinous "egg" that forms just beneath the soil surface. As the fungus matures, the egg ruptures, and the fingers emerge and rapidly expand. The fingers are initially white but quickly turn a vivid red or orange as they elongate. This transformation can happen within a few hours, making it a dramatic and sudden appearance in the landscape.

Odor and Spore Dispersal: The foul odor emitted by Clathrus archeri is often compared to rotting meat or decaying organic matter. This smell serves a crucial role in the fungus's reproductive strategy by attracting flies and other insects. These insects are drawn to the odor, mistaking it for a food source or decaying carcass. As they land on the slimy surface to feed, they inadvertently pick up spores, which are then transported to new locations when the insects fly away, facilitating the spread of the fungus.

Habitat: Clathrus archeri typically grows in leaf litter, mulch, or decaying wood in moist, shaded environments. It is commonly found in gardens, forests, and other areas where organic material is abundant. The fungus can thrive in a variety of climates but prefers temperate regions with adequate moisture.

Ecological Role: As a saprophytic fungus, Clathrus archeri plays an important role in the decomposition of organic matter. By breaking down leaf litter, wood, and other plant debris, it contributes to nutrient cycling and soil formation. The presence of Clathrus archeri in an ecosystem indicates a healthy process of organic matter decomposition.

The unusual appearance and odor of Clathrus archeri have led to various cultural associations and myths. In some cultures, the fungus is associated with folklore and superstitions, often viewed as an omen or a symbol of decay and death due to its smell and appearance. Its dramatic and somewhat unsettling emergence from the ground has earned it a place in stories and legends about supernatural phenomena.

Despite its off-putting smell, Clathrus archeri is harmless to humans and animals. However, its slimy spore-bearing substance can be unpleasant to touch and difficult to wash off. While it is not considered edible, there are no known toxic effects associated with the fungus.

Scientific research on Clathrus archeri and related stinkhorn fungi provides valuable insights into fungal ecology, reproduction, and evolutionary biology. Studying the mechanisms of spore dispersal, chemical composition of odors, and interactions with insect vectors helps scientists understand the complex relationships between fungi and their environments.

In conclusion, Clathrus archeri, the Devil's Fingers or octopus stinkhorn, is a remarkable and bizarre fungus that captivates with its striking appearance and foul odor. Its unique lifecycle, ecological role, and cultural significance make it a fascinating subject for study and observation. As both a symbol of the eerie and an important decomposer in its ecosystem, Clathrus archeri highlights the incredible diversity and adaptability of fungi in the natural world.

6. Nepenthes attenboroughii

Nepenthes attenboroughii, named after the renowned naturalist Sir David Attenborough, is one of the most fascinating and unusual carnivorous plants in the world. Discovered in the remote mountains of Palawan in the Philippines, this giant pitcher plant is known for its enormous, pitcher-shaped traps that can capture and digest small animals, including insects and even small mammals. The unique characteristics and life cycle of Nepenthes attenboroughii are as intriguing as they are bizarre:

Appearance: Nepenthes attenboroughii is a large, robust pitcher plant with distinctive, tubular traps that can reach up to 30 centimeters (12 inches) in height and 16 centimeters (6 inches) in diameter. The pitchers are green to reddish-purple and are equipped with a lid to prevent rainwater from diluting the digestive fluids inside. The inner walls of the pitchers are slippery, causing prey to fall into the trap and become submerged in the digestive enzymes at the bottom.

Carnivorous Mechanism: The plant's pitchers serve as pitfall traps, a common strategy among carnivorous plants to capture prey. Insects and small animals are attracted to the pitchers by nectar secreted along the rim and inner surfaces. Once inside, the prey finds it difficult to escape due to the slippery walls and downward-facing hairs. The digestive fluids at the bottom of the pitcher contain enzymes and bacteria that break down the prey into nutrients, which are then absorbed by the plant.

Habitat: Nepenthes attenboroughii is found in isolated, high-altitude regions of the Victoria Massif in Palawan, at elevations of around 1,600 meters (5,250 feet) above sea level. The plant thrives in nutrient-poor, acidic soils typical of these montane environments. The challenging growing conditions and the plant's dependence on capturing prey for nutrients highlight its remarkable adaptations.

Conservation Status: Due to its restricted habitat and the pressures of habitat destruction and climate change, Nepenthes attenboroughii is considered endangered. Conservation efforts are essential to protect this rare species and its unique ecosystem. In-situ conservation, such as protecting its natural habitat, and ex-situ conservation, including cultivation in botanical gardens, are both critical strategies for ensuring its survival.

Scientific Significance: Nepenthes attenboroughii, like other carnivorous plants, provides valuable insights into evolutionary biology and plant adaptation. Studying its mechanisms of prey capture, digestion, and nutrient absorption helps scientists understand the complexities of plant evolution in nutrient-poor environments. Additionally, the plant's discovery and subsequent research have highlighted the importance of preserving biodiversity and exploring remote.

7. Mycena chlorophos

Mycena chlorophos, commonly known as the glowing mushroom, is a captivating species of bioluminescent fungi. Native to subtropical regions such as Japan, Taiwan, Indonesia, and Brazil, this small mushroom is notable for its ability to emit a soft, greenish light in the dark. The phenomenon of bioluminescence, while rare in fungi, adds an otherworldly charm to Mycena chlorophos, making it a subject of fascination for mycologists and nature enthusiasts alike.

The unique characteristics and life cycle of Mycena chlorophos are as fascinating as they are enigmatic:

Appearance: Mycena chlorophos produces small, delicate mushrooms with caps that typically measure between 1 to 2

228

centimeters in diameter. The caps are pale and translucent, often with a slight greenish tint. The stems are thin and fragile, supporting the small caps. The mushrooms grow in clusters on decaying wood, leaf litter, and other organic matter in damp, shaded environments.

Bioluminescence: The most remarkable feature of Mycena chlorophos is its bioluminescence. The mushrooms emit a faint, greenish glow that is most visible in complete darkness. This light is produced by a chemical reaction involving luciferin (a light-emitting compound), luciferase (an enzyme), and oxygen. The exact biological function of bioluminescence in Mycena chlorophos is not fully understood, but it is believed to play a role in spore dispersal by attracting insects and other organisms that help spread the spores.

Habitat: Mycena chlorophos thrives in humid, subtropical climates. It is commonly found in forested areas with abundant decaying organic matter, such as fallen logs and leaf litter. The fungus prefers dark, damp environments, where it can flourish and produce its bioluminescent fruiting bodies.

Life Cycle: Like other fungi, Mycena chlorophos has a life cycle that includes both asexual and sexual reproduction. The mushrooms are the fruiting bodies of the fungus, responsible for producing and releasing spores. These spores are dispersed by wind, water, and animals, eventually landing on suitable substrates where they germinate and grow into new fungal colonies. The bioluminescence is typically most prominent during the spore-producing stage.

Ecological Role: Mycena chlorophos plays an important role in forest ecosystems as a decomposer. By breaking down dead organic matter, such as wood and leaves, the fungus recycles nutrients back into the soil, supporting the growth of plants and other organisms. Its bioluminescence, while a striking visual

229

phenomenon, also contributes to the ecological interactions within its habitat.

Scientific Research: The study of bioluminescence in Mycena chlorophos and other fungi has significant scientific implications. Understanding the biochemical pathways and genetic mechanisms underlying bioluminescence can lead to advancements in various fields, including molecular biology, biotechnology, and medical research. Bioluminescent fungi are also used in environmental monitoring, as their light-emitting properties can indicate the presence of certain pollutants or changes in environmental conditions.

Cultural Significance: Bioluminescent fungi like Mycena chlorophos have captured the human imagination for centuries. In some cultures, they are associated with folklore and myths, often seen as magical or supernatural. The enchanting glow of these mushrooms has inspired art, literature, and even modern scientific exploration, highlighting the enduring fascination with natural phenomena that blur the lines between reality and fantasy.

Conservation efforts are essential to protect the habitats where Mycena chlorophos and other bioluminescent fungi thrive. Forest conservation, sustainable land management practices, and efforts to reduce pollution and habitat destruction are crucial for maintaining the delicate ecosystems that support these unique organisms.

In conclusion, Mycena chlorophos, the glowing mushroom, is a remarkable example of nature's bioluminescent wonders. Its delicate appearance, combined with its ability to emit a soft, greenish light in the dark, makes it a captivating subject for study and observation. As a vital decomposer in forest ecosystems and a source of scientific inspiration, Mycena chlorophos highlights the incredible diversity and complexity of fungi and their roles in the natural world. Efforts to conserve its habitat and further

understand its bioluminescent mechanisms continue to shed light
on the many mysteries of this extraordinary fungus.

8.   Puya chilensis

Puya chilensis, a member of the Bromeliaceae family, is a peculiar
and formidable plant native to the arid hillsides of central Chile.
Known for its towering, spiked inflorescence and the peculiar claim
that it can trap animals, Puya chilensis has garnered interest from
botanists and plant enthusiasts around the world. The plant's
unusual adaptations to its harsh environment and its striking
appearance make it a noteworthy subject of study.

The characteristics and life cycle of Puya chilensis are as fascinating
as they are unique:

Appearance: Puya chilensis is a large, terrestrial bromeliad that can
grow up to 3 meters (10 feet) tall, with a spread of up to 2 meters
(6.5 feet). The plant forms a dense rosette of stiff, spiny leaves that
can be quite sharp, deterring herbivores from browsing. The leaves
are grayish-green and covered in small, sharp spines, making the
plant difficult to handle or approach.

Inflorescence: One of the most striking features of Puya chilensis
is its tall, flower spike, which can reach heights of up to 4 meters
(13 feet). The inflorescence consists of a dense cluster of bright
yellow-green flowers that attract pollinators such as birds and
insects. The flowering spike is covered with spiny bracts, adding to
the plant's formidable appearance.

Animal Trapping: Puya chilensis has gained a reputation for
trapping animals, particularly small mammals such as sheep. The
dense, spiny rosettes can entangle animals that attempt to feed on
the plant or accidentally brush against it. Trapped animals may
eventually die and decompose, providing the plant with nutrients.
While this behavior has led to sensational claims that the plant is

231

carnivorous, it is more accurate to describe it as a form of nutrient acquisition from decomposing matter, rather than active predation. Habitat: Puya chilensis is well-adapted to the arid conditions of its native habitat in central Chile. It thrives on rocky, sun-exposed hillsides where water is scarce. The plant's adaptations, such as its spiny leaves and ability to absorb nutrients from decomposing matter, allow it to survive in nutrient-poor soils.

Reproduction: The plant reproduces both sexually and vegetatively. The bright flowers are pollinated by birds and insects, leading to the production of seeds. These seeds are dispersed by wind or animals, allowing new plants to establish in suitable locations. Additionally, Puya chilensis can propagate vegetatively through offsets, which grow from the base of the parent plant and eventually form new rosettes.

Ecological Role: In its native ecosystem, Puya chilensis plays a role in providing habitat and food for various organisms. The dense rosettes offer shelter for small animals, while the flowers provide nectar for pollinators. The plant's ability to trap and decompose animals contributes to nutrient cycling in the arid environment.

Conservation efforts are important for protecting Puya chilensis, especially as its native habitat faces threats from human activities such as agriculture, urbanization, and climate change. Preserving the natural landscapes where this plant thrives is crucial for maintaining the biodiversity and ecological balance of the region.

In addition to its ecological significance, Puya chilensis has cultural and horticultural value. The plant's dramatic appearance and unique adaptations have made it a popular subject for botanical gardens and plant collectors. Its resilience and striking inflorescence make it an interesting addition to arid and xeriscape gardens, where it can thrive with minimal water.

Scientific research on Puya chilensis continues to shed light on the plant's adaptations, ecological interactions, and potential uses in sustainable landscaping and agriculture. Understanding how this plant survives in harsh conditions can provide insights into drought-resistant plant strategies and inform conservation efforts for other species in similar environments.

In conclusion, Puya chilensis is a remarkable and formidable plant that stands out for its towering inflorescence, spiny leaves, and unique adaptations to arid conditions. Its ability to trap animals and absorb nutrients from decomposing matter adds to its mystique and highlights the diverse strategies plants use to survive in challenging environments. As both a symbol of resilience and a subject of scientific interest, Puya chilensis underscores the incredible diversity and ingenuity of the plant kingdom.

9. Armillaria ostoyae

Armillaria ostoyae, commonly known as the honey fungus or the humongous fungus, is one of the most extraordinary and colossal fungi in the world. This species of fungus is infamous for its massive underground networks and its ability to cause root rot in a wide variety of trees and woody plants. Found primarily in temperate regions of the Northern Hemisphere, Armillaria ostoyae has captured the attention of mycologists and ecologists due to its incredible size and longevity.

The characteristics and life cycle of Armillaria ostoyae are as fascinating as they are complex:

Appearance: The visible fruiting bodies of Armillaria ostoyae, known as honey mushrooms, typically appear in the autumn. These mushrooms are characterized by their golden-yellow to brown caps, white gills, and a distinctive ring around the stem. While the mushrooms themselves are relatively small, the true extent of the fungus lies hidden beneath the soil.

Mycelial Network: Armillaria ostoyae is primarily known for its extensive mycelial network, which consists of thread-like filaments

233

called hyphae. These hyphae form a vast underground network that can spread over several acres and penetrate deep into the soil. The largest known Armillaria ostoyae network, located in the Malheur National Forest in Oregon, covers an area of approximately 2,385 acres (965 hectares) and is estimated to be over 2,400 years old, making it one of the largest and oldest living organisms on Earth.

Pathogenicity: Armillaria ostoyae is a pathogenic fungus that causes root rot in trees and other woody plants. The fungus infects the roots and lower stems, disrupting the flow of water and nutrients and ultimately leading to the death of the host plant. Infected trees often display symptoms such as yellowing leaves, reduced growth, and the presence of white mycelial fans under the bark. The fungus can spread from one tree to another through root-to-root contact or via rhizomorphs, which are root-like structures that facilitate the spread of the fungus through the soil.

Reproduction: Armillaria ostoyae reproduces both sexually and asexually. The honey mushrooms produce spores, which are dispersed by wind and can germinate to form new mycelial networks. Asexually, the fungus can spread vegetatively through the growth of its mycelium and rhizomorphs. The ability to reproduce both sexually and asexually contributes to the fungus's widespread distribution and persistence.

Ecological Role: Despite its pathogenic nature, Armillaria ostoyae plays a significant role in forest ecosystems. The fungus contributes to the decomposition of dead and dying trees, recycling nutrients back into the soil and supporting the growth of other plants. In healthy forests, the presence of Armillaria can help maintain ecological balance by promoting the turnover of plant species.

Control and Management: Managing Armillaria root rot is challenging due to the fungus's extensive underground networks and its ability to survive in the soil for many years. Control measures include removing infected trees and roots, improving soil

234

drainage, and planting resistant tree species. Biological control methods, such as the use of antagonistic fungi, are also being explored as potential strategies for managing Armillaria infections. Armillaria ostoyae holds significant scientific interest due to its unique biology and ecological impact. Research on this fungus provides insights into fungal physiology, plant-pathogen interactions, and the dynamics of forest ecosystems. Understanding the mechanisms of Armillaria's pathogenicity and its interactions with host plants can inform forest management practices and contribute to the development of more effective control measures.

In addition to its scientific importance, Armillaria ostoyae has captured the public's imagination due to its colossal size and longevity. The concept of a single organism spreading over hundreds of acres and living for thousands of years challenges our understanding of life and scale in the natural world. The "humongous fungus" serves as a reminder of the hidden complexity and interconnectedness of ecosystems, where even the smallest organisms can have a profound impact.

In conclusion, Armillaria ostoyae, the honey fungus or humongous fungus, is a remarkable and colossal organism that stands out for its extensive mycelial networks, pathogenic nature, and ecological significance. As both a forest pathogen and a decomposer, Armillaria plays a complex role in forest ecosystems, highlighting the intricate relationships between fungi, plants, and the environment. The study and management of this extraordinary fungus continue to provide valuable insights into the world of mycology and the broader dynamics of forest health and sustainability.

10. Dracaena cinnabari (Dragon Blood Tree)

Dracaena cinnabari, commonly known as the Dragon Blood Tree, is one of the most iconic and visually striking plants in the world. Native to the Socotra archipelago in Yemen, this unique tree is known for its umbrella-like canopy and the crimson resin it produces, which has been historically referred to as "dragon's blood." The Dragon Blood Tree's peculiar appearance and fascinating ecological adaptations make it a subject of interest for botanists and nature enthusiasts alike.

The distinctive characteristics and ecological significance of Dracaena cinnabari are as intriguing as they are unique:

Appearance: The Dragon Blood Tree is instantly recognizable by its dense, umbrella-shaped canopy, supported by a sturdy trunk. The tree can reach heights of up to 10 meters (33 feet). The leaves are clustered at the ends of the branches, giving the canopy its characteristic appearance. The bark is grey and smooth, often with scars from resin harvesting.

Dragon's Blood Resin: One of the most remarkable features of Dracaena cinnabari is the red resin it produces, known as dragon's blood. This resin is exuded when the bark or branches are cut, and it has been used for centuries for various purposes, including medicine, dye, incense, and varnish. In traditional medicine, dragon's blood has been used for its supposed healing properties, such as treating wounds and inflammation.

Habitat and Adaptations: The Dragon Blood Tree is adapted to the harsh, arid conditions of Socotra, where rainfall is scarce, and temperatures can be extreme. The tree's umbrella-shaped canopy is an adaptation to reduce water loss by providing shade to its roots and limiting evaporation. The thick, waxy leaves help conserve moisture, and the tree's extensive root system allows it to tap into deep water sources.

Reproduction: Dracaena cinnabari reproduces through seeds, which are dispersed by birds and other animals that feed on its fruit. The tree flowers once a year, producing small, fragrant white or greenish flowers that develop into fleshy berries. The seeds within the berries are capable of germinating under the right conditions, although the tree's slow growth rate means that it takes many years to reach maturity.

Ecological Role: The Dragon Blood Tree plays a vital role in its ecosystem. Its dense canopy provides habitat and shade for various plants and animals, helping to maintain biodiversity in the harsh environment of Socotra. The tree also helps prevent soil erosion with its extensive root system, contributing to the stability of the island's fragile landscapes.

Conservation Status: Dracaena cinnabari is classified as vulnerable due to habitat loss, overharvesting of resin, and the effects of climate change. The unique environment of Socotra is under threat from human activities, including agriculture and infrastructure development. Conservation efforts are crucial to protect this iconic species and its habitat. Initiatives include the establishment of protected areas, sustainable harvesting practices, and reforestation projects.

Cultural Significance: The Dragon Blood Tree holds significant cultural and historical value for the people of Socotra and beyond. Its resin has been traded and used for millennia, and the tree itself is an important symbol of the island's natural heritage. Local traditions and folklore often feature the Dragon Blood Tree, highlighting its importance in the cultural landscape of Socotra.

Scientific Research: Ongoing research on Dracaena cinnabari aims to understand its unique adaptations, reproductive biology, and ecological interactions. Scientists are also studying the chemical composition of dragon's blood resin to explore its potential medicinal properties and applications. Conservation biology research is focused on developing strategies to protect and restore populations of the Dragon Blood Tree in the wild.

In conclusion, Dracaena cinnabari, the Dragon Blood Tree, is a remarkable and iconic plant that stands out for its unique appearance, ecological significance, and cultural importance. Its ability to thrive in the harsh conditions of Socotra is a testament to the resilience and adaptability of nature. As both a subject of scientific inquiry and a symbol of natural beauty, the Dragon Blood Tree underscores the importance of conserving the world's unique and fragile ecosystems.